CAMBRIDGE LIBRARY COLLECTION

Books of enduring scholarly value

Technology

The focus of this series is engineering, broadly construed. It covers technological innovation from a range of periods and cultures, but centres on the technological achievements of the industrial era in the West, particularly in the nineteenth century, as understood by their contemporaries. Infrastructure is one major focus, covering the building of railways and canals, bridges and tunnels, land drainage, the laying of submarine cables, and the construction of docks and lighthouses. Other key topics include developments in industrial and manufacturing fields such as mining technology, the production of iron and steel, the use of steam power, and chemical processes such as photography and textile dyes.

Gold Mines of the Rand

The mining engineer and petrologist Frederick Henry Hatch (1864–1932) left the Geological Survey of Great Britain in 1892, relocating to South Africa. He worked for De Beers and with John Hays Hammond for Cecil Rhodes, finding important new gold fields in Matabeleland and Mashonaland. Control of the gold mines was a significant factor in the tension between Dutch and English settlers that would result in the Second Boer War in 1899. Prior to this, Rhodes and Hammond were behind the abortive Jameson Raid, but Hatch had returned to England briefly and was not implicated. This 1895 work, written with South African mining engineer J.A. Chalmers, reveals the extent of gold reserves in the Transvaal, and the engineering skills needed to exploit them. It deals with geological, economic and legal aspects of the mining industry, remaining of interest to historians of South Africa and the British Empire.

Gold Mines of the Rand

Being a Description of the Mining Industry of Witwatersrand, South African Republic

F.H. Hatch
J.A. Chalmers

Cambridge
University Press

CAMBRIDGE UNIVERSITY PRESS

Cambridge, New York, Melbourne, Madrid, Cape Town,
Singapore, São Paolo, Delhi, Mexico City

Published in the United States of America by Cambridge University Press, New York

www.cambridge.org
Information on this title: www.cambridge.org/9781108061667

This edition first published 1895
This digitally printed version 2013

ISBN 978-1-108-06166-7 Paperback

The original edition of this book contains a number of colour plates, which have been reproduced in black and white. Colour versions of these images can be found online at www.cambridge.org/9781108061667

THE GOLD MINES OF THE RAND

THE

GOLD MINES OF THE RAND

BEING A DESCRIPTION OF THE MINING INDUSTRY OF WITWATERSRAND SOUTH AFRICAN REPUBLIC

BY

FREDERICK H. HATCH
(MINING ENGINEER)

AND

J. A. CHALMERS
(MINING ENGINEER)

WITH MAPS, PLANS, AND ILLUSTRATIONS

London
MACMILLAN AND CO.
AND NEW YORK
1895

PREFACE

At the present moment the Transvaal is undoubtedly one of the most interesting countries in the world.

The enormous wealth which lies buried in the Witwatersrand Gold Fields has attracted capital, enterprise, and talent, the three factors essential to the proper development of a country. With marvellous rapidity a great mining industry has sprung up, the ultimate limits of which it is difficult to assign; and the recognition of the fact that it will be possible to work the ore-deposits down to great depths, is giving rise to engineering problems of the greatest moment.

The regularity of their geological formation has ensured for these auriferous beds a continuance in length and depth, which is unique in the history of Gold Mining. This fact, combined with the success achieved by the outcrop properties, has won for the fields the confidence of capitalists to a degree remarkable in mining, and has induced the outlay of the large sums of money which have been, and still are, requisite to fully develop their resources.

But notwithstanding the fact that the keen interest in the Fields evinced by the public shows no signs of abating, but appears rather to be on the increase, the available published information relating to the nature of the ore-deposits and to the extraction of the Gold on the Rand has remained meagre and inadequate.

It is certain that a book dealing adequately with this subject must be of the greatest service to those interested in the Industry, and this idea suggested the present work.

In writing it, I have had the advantage of the collaboration of my friend, Mr. J. A. Chalmers. A long residence on the fields and the opportunities incidental to the pursuit of our profession as mining engineers have given us unusual facilities for obtaining the necessary facts; but we are also indebted to many professional gentlemen on the Rand for valuable information, and I take this opportunity of thanking them. Interesting particulars relating to the Cyanide process of gold extraction have also been contributed, and appear in the chapter devoted to that section of the subject.

The excellent map of the Witwatersrand mining properties has been prepared at my request by Mr. Melvill.

I am specially indebted to Mr. John Hays Hammond (with whom it has been my good fortune to be closely associated in much of my work in South Africa) for the clear exposition of his views on the future of this Industry, which I append hereto. Mr. Hammond's position in the Mining World is too well known to need any comment on my part.

FREDK. H. HATCH.

LONDON, 30*th July* 1895.

Mr. Hammond writes:—

THE FUTURE OF THE RAND

As to the immediate future of the Rand, that is, the future of the Rand so far as the lives of the outcrop companies are concerned, there can be but one opinion, and that of a decidedly favourable tenor. If any reassurance were needed, this important work of Messrs. Hatch and Chalmers would, I think, be "confirmation strong as proofs of Holy Writ."

Just how many years are comprehended in the immediate future of the Rand is not a matter of easy determination, but

I am inclined to the opinion that the majority of engineers have been too conservative in their estimates, for we have many instances here of the paradox of the lives of mines increasing *pari passu* as exploitations progress. This circumstance is due to the fact that, as the economic conditions of mining become more favourable, and as the mines are operated upon a larger scale, and with better management, lower-grade ore becomes more profitably mined, and the available tonnage of payable ore correspondingly increased. The number and life of the outcrop mines of the Rand will be extended through the discovery of payable reefs, in what are known as outside districts, where investigations have shown the existence of well-defined banket reefs. But it is to the development of the Deep-Level areas that the Witwatersrand district must look for the most important augmentation of its life; and it is to the subject of deep-level mining, and more particularly the mining of the deep-level areas embraced within what is known as the central section of the Rand, that I confine my discussion of the future of the district.

As is well known, the Mining Laws of the Transvaal do not give the claim-holder the right of the lateral pursuit of his reef in depth, but confine him to the ground lying vertically below his surface boundaries. The claims are parallelograms of 150 Cape feet in the direction of the strike, and 400 Cape feet in that of the dip of the reefs. The claims first pegged were located as near as possible to the outcrop, or to the apex of the reef; and what are known as "outcrop" companies were formed to exploit an amalgamated block of contiguous claims, since a single claim was obviously too small to admit of profitable mining.

As the result of developments upon the outcrop companies, and the bore-hole tests, demonstrated the value of the area lying to the dip of the reefs, what were called deep-level companies were formed to exploit such ground. The term deep level as applied to these mines is a misnomer, inasmuch as very few of the shafts to reach the reefs exceed 1000 feet in vertical depth—a shallow depth compared with mining experience in other countries. Relatively, however, the term is a convenient one when applied in contradistinction to the outcrop companies.

As the geological character of the Rand deposit became known, and as the remarkable persistency of the reefs in depth, and the unique uniformity in the gold-tenure of the banket became established upon the deep-level properties, and by still deeper bore-holes, the potentialities of the deep-level areas, situated below the first row of deep levels, were recognised, and companies were organised to work what may be called the second row of deep levels. These companies are, properly speaking, the deep-level companies. But even the term deep level is here merely relative, for the depths of the projected shafts to work the areas embraced by these companies are still shallow, compared with the deep-level shafts of America and Europe, and the term should therefore be accepted with this qualification.

The feasibility of successful mining upon the area comprised by the outcrop and first row of deep-level mining companies is practically assured. It is to the mining of the deeper areas embraced within the second row of deep-level claims that I confine my remarks. The problem is one of great importance, not only to the Transvaal, but to the world in general; for on its successful solution depends the gain to the world of an enormous gold output, as appears, if we accept the conservative estimate of £52,000 as the yield in gold per claim. Upon this basis it will be seen that every 1000 feet of vertical length, assuming the reef to be 30° in its dip, would open up ore from which £260,000 of gold would be obtained for every 150 feet, the length of one claim along the course of the reef. Every mile of ground parallel to the course of the reefs would therefore yield per 1000 feet of vertical depth the sum of about £9,000,000. The problem presented then is, to how many thousands of feet in depth can mining operations be carried in the Witwatersrand mining district? From an engineering point of view, mining operations are certainly practicable to a vertical depth of at least 5000 feet. There is no necessity for abstract reasoning upon this point, as the practicability of mining to these depths has been demonstrated in other countries. But regarded from a financial point of view, will mining to this depth pay? The initial expense, as compared with that of mining upon the outcrop companies, will undoubtedly be considerably greater on account of the increased cost of the heavier plant required, and of the shaft sinking involved; therefore it will be necessary in order to ensure profitable mining upon such deep-level properties, that larger areas be embraced in one company, so as to minimise the capital expenditure per claim worked. If we work, for example, by means of deep-level plants, blocks of 250 claims to the depth of 5000 feet, the extra cost per claim, as compared with the cost of the plants to work the outcrop companies, will not be more than £1500 a claim—surely not a prohibitive expenditure, if we bear in mind the fact that the profits per claim upon the central section are estimated at upwards of £18,000. As a matter of fact, the capital expenditure through faulty engineering in the case of the outcrop companies, has been considerably greater than the estimate I have allowed in this comparison of the relative costs of mining the outcrop and deep-level companies respectively. The cost of mining would likewise undoubtedly increase, compared with that of the most economically managed of the outcrop companies; but in my judgment these increased expenses would not, even under present economic conditions, greatly exceed those of the average mining costs of the outcrop companies. The increased cost would be due chiefly to the larger expenses attending hauling, pumping, and ventilation; but even under present conditions as to cost of labour and supplies, the extra costs of deep-level mining to a depth of 5000 feet should not exceed five shillings per ton; and this is obviously not an insuperable obstacle to profitable mining. In my opinion there will be even less disparity of cost than this, owing to improved economies that would be effected by conducting the operations upon a larger scale than customary, by obtaining cheaper, or in any event what is equivalent, more efficient labour, both white and black, and by cheaper supplies. Therefore, looking to the time when deep-level mining will be carried on extensively, we may reasonably predict such betterment of the economic conditions as to admit the mining of ore at the same working costs per ton as now obtain in the outcrop companies. The successful working of these deep-level areas presupposes the persistency of the conditions of the banket formation as to the widths and gold contents of the reefs. That such conditions obtain is evidenced by the geological character of the formation, and confirmed by the facts elicited through the actual workings upon the first row of deep-level claims, also from the deep-level bore-holes.

That poor zones are liable to be encountered, as in the case of the outcrop companies, is admitted; but, from a mining point of view, there can be little doubt that the reefs in the second row of deep levels will be equally good, both as to regularity in size and good values, as those developed upon the first row. In my opinion there is less uncertainty as to the mining value of the ground of the second row of deep levels than was entertained before mining operations were undertaken upon the first row. Between the first row of deep-level properties, and the 5000 feet vertical depth, to which depth at least I regard mining practicable, there lies a valuable auriferous area, which within a few years will be the scene of profitable mining operations.

The output of the Witwatersrand district for the year 1894 was 2,024,163 ounces of gold, with a value of £6,963,100.

That the output would be greatly increased for the year 1895 is indeed already assured, and it is equally certain that this year's output will place the Rand *facile princeps* of the gold-mining countries of the world; but even here the progress will not halt, and it is safe to predict that there will be a progressive increment of the annual output for many years to come. It is obviously impossible to make an approximate estimate, but I would regard as well within the bounds of conservatism the prediction that the annual output before the end of the present century will exceed £20,000,000 sterling worth of gold.

JOHN HAYS HAMMOND.

CONTENTS

CHAPTER VIII

CHAPTER IX

CHAPTER X

CHAPTER XI

CHAPTER XII

LIST OF ILLUSTRATIONS

LIST OF PLATES

ERRATA

For The Gold Mines of the Rand

p. 9, footnote 3, add *Vol. 44, 1888, p. 239.*
p. 18, line 3, for *Rooderval* read *Roodeval.*
p. 20, line 15, for *Verdefort* read *Vredefort.*
p. 84, footnote, line 4, for *of* read *by.*
p. 109, heading of second column of table, for *on* read *in.*
p. 227, heading of Section 2, for *precipitations* read *precipitation.*
p. 239, fourth par., the quotation marks should be placed after the word *treatment* in the fifth line.

CHAPTER I

INTRODUCTION

BEFORE the year 1868 the Boers, wishing to discourage the inroad of foreigners, allowed no digging for gold within the boundaries of the South African Republic. In that year, in consequence of the successive discoveries of auriferous quartz-veins and alluvial deposits, President Pretorius, more advanced in ideas than his predecessors, and impelled by the growing poverty of the State, succeeded in effecting the repeal of the laws against prospecting, and rewards were even offered for the discovery of payable gold. As a result of these changes gold was discovered, in 1869, by Edward Button in the Sutherland Hills, in the Kleinletaba, and, in 1870, in the Murchison Range. In 1871 a mining commissioner was appointed in Zoutspansberg, and the first gold law was promulgated. In 1876 the Mac Mac and Pilgrim's Creek alluvial deposits were exploited, and some work was done in the Lydenberg district during the years succeeding the war of Independence of 1881.[1] The De Kaap Gold Fields were discovered in 1884, and the Sheba Company was floated in 1885, with a capital of £15,000. The Witwatersrand deposits were found in 1885, by a man named Arnold, working on the farm Langlaagte,

[1] Since then these fields have not been so productive as at first, but gold is still won, though more from quartz-veins than from alluvial deposits. Within the last few months, however, there have been signs of a renewal of activity in the district.

and by December of the same year a five-stamp battery had been erected by the Brothers Struben. During the succeeding year the outcrop of the conglomerate beds was traced and found to be auriferous for many miles. On the 18th July 1886 the farms Roodepoort, Vogelstruisfontein, Paardekraal, Langlaagte, Turffontein, Randfontein, Doornfontein, Elandsfontein and Driefontein, were proclaimed as public diggings. The usual rush followed, and soon what had hitherto been an unattractive stretch of veldt became a scene of life and activity, and the value of ground, both mining and residential, rose rapidly.

The township of Johannesburg was marked off towards the end of 1886. The first sale of building sites took place in December 1886, and realised £13,000, prices ranging from a few shillings to £200. The first mill of any size was a 30-head Frazer and Chalmers battery, erected early in the year 1887 for the Paarl-Pretoria Gold Mining Company.

Coal was discovered in December 1887 at Boksburg in the near neighbourhood of the diggings, and this discovery led to the opening of several collieries (Brakpan, Springs, Olifants River, etc.). There is no doubt that the prosperity of the Rand Gold Fields is in a measure due to the fortunate accessibility of abundant and, consequently, cheap fuel.

Shares began to be quoted on the Johannesburg Stock Exchange in June 1887. By November of the same year sixty-eight companies had been formed with a nominal capital of £3,000,000. A strong upward movement in the price of shares took place in the following July and August, and this culminated in a remarkable period of activity and excitement known as the boom, which started in November 1888, and lasted with fluctuations to the end of January 1889. There was then a lull, very little business being transacted. In March a heavy fall took place and continued till the autumn, when, after a slight rally, a panic ensued, consequent on the banks stopping credit; and the market rapidly fell to pieces. This boom was characterised by the usual features of such movements : — unlimited credit, enormously overdrawn accounts at the banks, and the wildest speculation. It is difficult to gauge its precise effect on the development of the fields. Some are of the opinion that the Rand thereby received a world-wide advertisement causing money to be brought into the country, and facilitating the raising of working capital for developing properties ; many, on the other hand, believe that the loss of confidence that resulted, retarded development. In any case the fever of speculation having once burnt itself out, men had to turn to steady work to prove and open up the resources of their properties. There is no doubt that for several years Rand Gold Mines were neglected as a field for investment for foreign capital, the recovery of public confidence being no doubt checked to some extent by the general timidity of speculators and investors consequent on the Baring crisis of 1891. But the ever-increasing output and the augmentation of dividends in the years 1892 and 1893 began to attract the attention of other countries, resulting finally in the Russian Government sending out a special commissioner, Mr. Kitaeff, to report on the Witwatersrand fields, and in the German Government despatching Mr. Bergrath Schmeisser for the same purpose, while a number of private mining engineers of repute were sent out from France and England. The favourable reports of these engineers, coupled with the increased production and augmented dividends, brought about a great revival of interest at the close of last year (1894), which has led to large purchases of shares in Rand Gold Mines being made by capitalists and investors in France, Germany, and other European countries.[1]

The important influence exercised by the discovery of the Witwatersrand Gold Fields on the prosperity of the

[1] Reckoned at present Stock Exchange rates, the market value of the Witwatersrand mines amounts to over one hundred millions sterling.

Transvaal is shown by the marvellous expansion in trade and business which has since taken place. This is evidenced in the following table, which gives the revenue and expenditure of the State from 1882 to 1894 :—

Year.	Revenue.	Expenditure.
	£	£
1882	177,406	114,476
1883	143,323	184,343
1884	161,595	184,822
1885	177,876	162,708
1886	380,433	211,829
1887	668,433	621,073
1888	884,440	720,492
1889	1,577,445	1,201,135
1890	1,229,060	1,386,461
1891	967,191	1,350,073
1892	1,255,829	1,188,765
1893	1,702,684	
1894	2,247,728 [1]	

The rapid growth of the Witwatersrand fields is shown in the following table, which gives the annual output from mines working on conglomerate reefs in the area included under the head Witwatersrand :—[2]

Year.	Ounces.	Value.
		£
1887	23,149	81,022
1888	207,660	726,821
1889	369,557	1,300,509
1890	494,817	1,735,491
1891	729,238	2,556,328
1892	1,208,928	4,290,733
1893	1,476,502	5,180,090
1894	2,023,198	6,959,622
1895 (6 months)	1,121,761	3,839,923
	7,654,810	£26,670,539

[1] Of this sum the Official Mining Department contributed £972,311, or 43¼ per cent. *State Mining Engineer's Report* for 1894.

[2] This includes the district of Heidelberg, but not those of Klerksdorp and Potchefstroom.

Alluvial gold is not included. In the Chamber of Mines returns[1] an allowance of 42,000 ounces is made for the estimated unrecorded production during the years 1887, 1888, and 1889. This brings the total to 7,696,810 ounces, of a value of £26,817,484.

The total gold production of the Transvaal for the year 1894 is thus given by the Chamber of Mines :—[2]

Witwatersrand	2,024,163 ounces.
De Kaap	92,577 "
Lydenberg	60,275 "
Klerksdorp and Potchefstroom	77,714 "
Zoutpansberg	10,629 "
Malmani	494 "
Total	2,265,853 ounces,

having a money value estimated at £7,800,000.[3]

As the world's gold production for the year 1894 is estimated to have been £36,000,000, the amount contributed by the Transvaal is over one-fifth of the total.[4] The Transvaal ranked third in the list of producers for the year,

[1] *Report*, 1894, p. 240. [2] *Report* for 1894, p. 242.

[3] The Government returns are 2,239,865 ounces, valued at £7,667,152. According to the figures supplied to the Chamber of Mines by the Collectors of Customs of Cape Colony and Natal, 2,129,781 ounces of raw gold, of a value of £7,370,058, were exported from South Africa during the year 1894.

[4] The history of the world's gold production during the past half-century is briefly as follows :—In 1849 the total production only amounted to £6,000,000. In 1853 it rose suddenly to £30,000,000, on account of the immense alluvial finds in California and Australia. From 1853 there was a gradual decline, the lowest point being reached in 1883 with £20,000,000. From 1883 to 1887 there was a slow increase, the production being greatly aided by the output of the El Callao mine in Venezuela and the Mount Morgan mine in Australia. Since the Witwatersrand Mills commenced running in 1887, the production has advanced rapidly; the figures for the last five years are as follows :—

1890—£23,700,000
1891—£26,130,000
1892—£29,260,000
1893—£31,110,000
1894—£36,000,000

the United States being first with an output of £7,952,240, and Australasia second with £7,855,639.[1]

There is not the least doubt in the minds of those who know the conditions that prevail in the Witwatersrand Gold Fields that the production of the South African Republic will continue to show a vigorously-increasing growth for several years to come. On p. 287 we give the facts and figures on which we base our calculation that at the beginning of the next century the annual output from the Witwatersrand mines will have reached a total of 6,500,000 ounces, having a value of, say, £26,000,000. Various forecasts have been made of the total yield to be expected from the auriferous deposits of the Rand; and such estimates are of course influenced greatly by the attitude adopted by individual engineers in regard to the depth to which mining operations will ultimately be carried. Mr. Hamilton Smith places the limit of profitable mining somewhere between 3000 and 3500 feet, and estimates a total yield of £325,000,000 for the central section of the Rand[2] (a distance of about 11½ miles, viz. between the Langlaagte Block B and the Glencairn). Bergrath Schmeisser appears to have arrived at much the same conclusions, for he puts the probable output at £349,000,000[3] for the same length of outcrop. On p. 105 we give our reasons for believing that there will be no insuperable difficulties, either physical or mechanical, to prevent the development of the mines down to 5000 feet. Assuming, however, a vertical depth of only 3500 to 4000 feet, or more exactly 8000 feet on the incline, we estimate that the total production to be expected from the main section of the Rand,

[1] The totals in 1893 were—

United States	£7,191,000
Australasia	6,560,000
Transvaal	5,480,497

[2] *Times*, London, 17th Jan. 1893 and 19th Feb. 1895.

[3] Schmeisser, *Ueber Vorkommen und Gewinnung der nutzbaren Mineralien in S. Afrik Republik*, Berlin, 1894, p. 149.

i.e. from Roodepoort to Driefontein (inclusive), a distance of 27 miles,[1] will amount to £592,000,000, or, if we include the outlying portions of the district, upwards of £700,000,000. These figures may at first sight appear extravagant, but as they are based on results actually obtained by the mining companies, and represent the logical deduction from the facts which have up to the present been rendered available by the developments of Witwatersrand mines, we are prepared to stand by them.

With regard to the duration or life of the Witwatersrand mines, either collectively or individually, we refrain from speculating, as we consider such estimates worse than useless, being dependent on so many indeterminate factors. It is evident that in any given mining area there is only a certain limited quantity of the noble metal, and exactly what proportion of it will be extracted must depend on the particular circumstances that govern the cost of obtaining it. These circumstances are constantly undergoing changes (*e.g.* reduction in the price of supplies and of labour, the use of improved and more economical plant); and such changes entirely alter the prospective life of a mine by bringing within the margin of profitable mining ore-bodies which before may have been considered too expensive to work. Again the life of a property necessarily depends on the stamping power employed. The total yield may be spread over 20 years with a fifty-stamp mill, or the mine may be worked out in 10 years with a hundred stamps, the same quantity of gold being ultimately obtained in both cases. It will assuredly be granted that it is a sound principle to obtain the greatest quantity of the metal in the shortest possible time

[1] There can be no object in excluding from the calculation ground on which there are already working mines like the Princess, Durban-Roodepoort, Roodepoort-United, Croesus, Knight's, Ginsberg, and the Comet, not to speak of many others that are rapidly coming to the fore. Our figure for the central portion of the Rand, from the Langlaagte Estate to the Glencairn, calculated on the same data, is £382,000,000.

and at the least possible cost, and these conditions favour a short life. By the opening up of two or three rows of deep-level properties as distinct propositions, without waiting for their development downwards from the outcrop, the annual production will be increased enormously; but since the total output must remain the same, the life of the Fields will be proportionately reduced.

With regard to profits, twenty-two companies paid dividends last year (1894), the sum distributed amounting to £1,489,307. During that year many of the companies spent a large proportion of their earnings in acquiring new plant and in the construction of additional works. This expenditure has placed them in a position to earn greater profits, both by increasing their production and by lowering the working costs. Already during the first six months of this year the sum of £1,094,467 has been paid in dividends by twenty-five companies, and there is every reason to anticipate that both the number of dividend-paying companies and the amount of profits to be distributed to shareholders will undergo in each succeeding year an augmentation proportionate to the increased output.

CHAPTER II

GEOLOGICAL

THE earliest writer on South African geology was Andrew Geddes Bain,[1] who in 1852 roughed out the chief features of the stratigraphy of the country. Since then other contributions have been made, notably by E. J. Dunn,[2] Professor A. H. Green,[3] Dr. A. Schenck,[4] W. H. Penning,[5] Walcot Gibson,[6] and Dr. Molengraaff.[7]

The first geological map of South Africa was made by E. J. Dunn.[8] Since then no one has attempted a geological map of South Africa, but several maps of the different Transvaal goldfields have been published,[9] some of which give particulars as to the occurrence of the auriferous deposits. The Government of the South African Republic has frequently been urged to establish an Official Geological Survey Department, and although the appointment of a State Geologist has on several occasions been mooted, hitherto nothing definite has been done in that direction. The importance of having the geological formations properly mapped, and the rapidly-

[1] *Trans. Geol. Soc.*, vii. pp. 175-192.

[2] *Geol. Mag.*, 1885, p. 171.

[3] *Quart. Journ. Geol. Soc.*

[4] *Peterm. Mitt.*, 34, 1888, p. 225, and *Zeit. Deutsch. Geol. Ges.*, 1889, p. 573.

[5] *Quart. Journ. Geol. Soc.*, 47, pp. 451-463, 1891.

[6] *Quart. Journ. Geol. Soc.*, 48, p. 404.

[7] *N. Jahrbuch f. Min.*, Beilage-Band, ix. 1894, p. 174.

[8] *Geological Sketch Map of South Africa*, London, 1887.

[9] By Troye, Melville, Jeppe, Maidment, and Goldmann.

accumulating information relating to the occurrence of the auriferous deposits collected and properly recorded, cannot be exaggerated; and the expenditure of a comparatively small sum of money in providing the means of doing this would be amply repaid by the benefit that the industry and the country generally would without doubt receive therefrom.

The main subdivisions of South African stratigraphy, as given by Schenck and Green, are, in descending order:—

Recent Deposits.
Karoo Formation.
Cape Formation.
South African Primary Formation.

The Primary Formation forms the crystalline floor on which rest the younger sedimentary deposits. Consisting of granites, gneisses, and crystalline schists, it constitutes the major portion of the formation of North-west Central Africa. It occurs largely in Mashonaland, Matabeleland, and the Mozambique, and predominates in the northern and eastern parts of the Transvaal, namely, in the districts of Zoutpansberg, the Murchison Range, Kleinletaba, Lydenberg, the Kaap Valley, and Swasiland. In the Witwatersrand district this ancient formation emerges from beneath the younger deposits between Pretoria and Johannesburg.

As a rule the schists are tilted to a high angle, and show evidence of having undergone considerable contortion; they consist in part of much-metamorphosed beds of sedimentary origin, such as slates, quartzites, and magnetite-quartz rock; but to a greater extent of hornblendic, chloritic, and serpentinous schists that have been derived by mechanical and mineralogical metamorphism from basic igneous rocks (diabases, diorites, and other greenstones).[1]

On account of the wide-spread occurrence of these beds in Swasiland Schenck names them Swasi-schists. He

[1] See a paper by the Authors on Mashonaland and Matabeleland, published in the *Geological Magazine*, London, May 1895, p. 193.

suggests a correlation with European Silurian rocks. The Malmesbury beds of other geologists belong to this formation.

The Cape Formation.—Lying unconformably on the basement beds are the shales, sandstones, conglomerates, and limestones of the Cape Formation. This name was given by Schenck, who regards the Witwatersrand beds as of the same age as the Table Mountain sandstone. Penning[1] uses the name Magaliesberg Beds; but as the stratigraphical position of the Magaliesberg quartzite is uncertain, this nomenclature is not advisable. These varied deposits extend over the southern, western, and middle parts of the Transvaal, including the Heidelberg, Potchefstroom, Marico, Rustenburg, Pretoria, and Lydenberg districts. The long range of the Drakensberg consists of these beds, and thus they probably extend to the Cape, and are represented by the grits and sandstones of Table Mountain, and by the shales, sandstones, and quartzites of the Bokkeveldt beds. The fossil evidence afforded by the latter beds indicates an age corresponding to the Devonian and Lower Carboniferous periods of European classification.

The Karoo Formation is subdivided as follows (descending order):—

Schenck.	*Green.*	
Stormberg beds (Molteno beds)	Stormberg beds	Volcanic beds. Cave sandstone. Red beds. Molteno beds.
Beaufort beds . . .		Karoo beds. Kimberley beds.
Ecca beds (including the Dwyka conglomerate) .		Ecca beds. Dwyka conglomerate.

This formation, which has a widespread occurrence in the Cape Colony, Natal, and the Free State (Karoo, Nieuwveldt, and Camdeboo Hill range, Stormberg, Molteno, Indwe, etc.),

[1] *Quart. Journ. Geol. Soc.*, xlvii. p. 452, 1891.

derives its importance for the Transvaal from the fact that it carries the coal seams which have rendered such valuable assistance to the development of the auriferous deposits. In the form of light-coloured coarse-grained sandstones belonging to the Stormberg beds, and lying flat on the upturned edges of the older Cape Formation, it covers extensive tracts of the Southern Transvaal, and, crossing the Vaal River, extends into the Free State and Natal.

The coal deposits are confined to the Molteno division of the Stormberg beds. The principal seams worked in the Colony are in the Stormberg range (*e.g.* at Molteno, Cyphergat, Bushman's Hoek, etc.). In Natal the chief workings are in the neighbourhood of Newcastle and Ladysmith. Extensive coal deposits exist in the Free State, but have not yet been exploited to any great extent. In the Transvaal an extensive and valuable coal seam is worked by Messrs. Lewis and Marks near Vereenigen, north of the Vaal River; the other principal collieries are those at Brakpan (Coal Trust), Daggafontein (Cassel Colliery), Boksburg, and in the Middelburg district.

With regard to the age of the Karoo formation, palæontological evidence indicates the possibility of a correlation with Lower Mesozoic formations (Trias, Permian, or Upper Carboniferous).

Recent Deposits.—These deposits comprise those of alluvial and æolian origin, together with the curious surface material known as laterite, which is formed in hot and dry countries by the decomposition of rock *in situ.* In the Transvaal, laterite is of wide-spread occurrence in the form of a ferruginous nodular substance known to the Boers as *oudklip* (*i.e.* old stone).

A considerable portion of the surface is deeply covered by the products of atmospheric decomposition and water-borne deposits of red clay, sand, and loam. Many an important reef outcrop is thus hidden from view by a thick layer of what the gold-digger terms "wash," a condition of things which tends to make the work of the prospector both difficult and expensive.

Having thus briefly sketched the main features of the geological structure of South Africa, we now propose to describe in greater detail the formation that carries the auriferous conglomerates of Witwatersrand.

A section taken across the outcrop of the Cape Formation as represented in this district gives the following sequence of strata. The beds resting immediately on the granite of the Primary Formation at Johannesburg forms the elevated tract known as the Witwatersrand, or, briefly, as the Rand. This range constitutes the division or water-parting between the streams flowing northwards to the Limpopo River, and those tributary to the Vaal River on the south. The strata consist of hard white quartzites alternating with bluish clay shales, the latter being highly impregnated with oxides of iron, and even carrying in places well-defined seams of specular iron or hematite. The series, which, following Gibson, we will briefly designate as the Quartzite - Shale Group, is a very characteristic and important geological feature, as it immediately underlies the auriferous conglomerates or "banket" beds of the Main Reef Series, and is invariably associated with them. Geotectonically it may be described as a range of low hills, in which the quartzites protrude as ridges following the strike of the beds, while shallow valleys or depressions mark the position of the shale beds. In places the hills grade imperceptibly into the rolling plains of the high veldt, and the beds are hidden from view by a deep reddish surface deposit, an occasional boulder or isolated outcrop alone betraying their presence. But in the neighbourhood of Johannesburg, and for several miles east and west of the town, the hills are a prominent feature of the scenery.

To the south of the Quartzite-Shale Formation, and overlying it, is a thick series of reddish sandstones and quartzites, which at intervals enclose conglomerate deposits of greater or less thickness, to which detailed reference will be made later on. These beds conform to the underlying group, having like the latter an east and west strike, with a dip to the south. They form a flat or rolling expanse of country slightly sloping to the south. About three miles south of the Rand hills the beds are capped by a hard fine-grained greenstone or melaphyre, which rises into the Klipriversberg Range. South of this again are more sandstones, and in these occur the highest conglomerate beds of the series, namely the Black Reef.

The dip of the beds at the outcrop of the bottom conglomerates (Main Reef Series) is high, varying as a rule from 45° to 60°, but in places it is even higher, while in a few instances it falls to 15° or 20°. Towards the south the dip shown at the outcrops rapidly decreases, and at the Black Reef the formation lies very nearly flat. It is probable, however, that there is an unconformity between the upper (Black Reef) beds and the underlying formation. We shall have occasion to refer to this later on.

To the south of the Black Reef lies a broad expanse of flats and swamps some 18 or 20 miles broad. Here the rocks are buried beneath deep surface deposits, and but little is known of the solid formation. These plains are terminated by the Heidelberg Range, a bold line of hills lying to the north of the town of Heidelberg and stretching east and west; and here again we find the Witwatersrand formation outcropping, first the upper sandstones without conglomerates, then the conglomerate beds themselves, the formation dipping to the north.

Turning now to the country north of Johannesburg, we find immediately north of Johannesburg a tract of country in which the ancient floor of granite and schistose rocks is exposed at the surface. Some 20 or 30 miles farther north, however, the beds of the Cape Formation are again encountered, and are found to dip to the north. These beds cannot be exactly correlated with those occurring south of the granite, and the conditions under which they were deposited must have been somewhat different. It is true that in the Witwatersberg and Magaliesberg Ranges we also have a quartzite-shale formation; but the beds differ from the quartzites and shales of the Rand, being more numerous and of greater thickness. Conglomerates are absent, or at least have not yet been found. Underlying the quartzites and shales of the Witwatersberg is a remarkable deposit of hard bluish dolomite or magnesian limestone, forming the rugged country known as Kalkheuvel or Lime Hills. This limestone rock is highly impregnated with silica (quartz, chert, chalcedony), and with various lime-magnesia silicates (tremolite, asbestos, etc.), in consequence of which the surface has a curious wrinkled or corrugated character, produced by the soft parts being worn away under the influence of the weather, while the more siliceous and cherty portions have become prominent as ribs and ridges. This characteristic appearance has earned for it the Boer name of Olifantsklip (Elephant-Rock) from a fancied resemblance to the hide of that animal. The same limestone has a wide distribution in the western Transvaal (Marico District, Malmani, Apjatar's Kop), and can be traced into Bechuanaland, where it extends over a considerable amount of country.

Summarising briefly, we find that, of the section of country described above, the portion lying south of the Rand Hills has the geotectonic character of a flat syncline or trough; while lying between the Rand Hills and the Witwatersberg and Magaliesberg Ranges is the part of a large saddle or anticline, of which the axial portion has been removed by denudation so as to expose the underlying Primary Formation. The beds of the Witwatersberg and Magaliesberg Ranges form the northern limb of this anticline. Whether the corresponding southern limb is to be sought in the Witwatersrand Hills, or is represented by the Gatsrand series of quartzites and shales, and by the belt of limestone which lies to the north of it, is a matter about which there is some diversity of opinion. According to the former view, which received the support of Schenck,[1] the dolomitic limestone underlies the conglomerate series. Molengraaff,[2] however, maintains the latter opinion, regarding the dolomite and Gatsrand Beds as the upper portion of the Cape Formation, and as identical with the northern limestone and Magaliesberg Beds. On this hypothesis the conglomerates might be expected to be found outcropping on the northern side of the anticline, unless cut out by a fault. The facts at hand, however, are not sufficient to definitely decide this important question at the present time.

The outcrop of the conglomerate beds has been followed and studied by quite a host of enterprising prospectors allured by the chances of fortune. In consequence of such close and eager study, an almost continuous stretch of outcrop, extending east and west of Johannesburg over nearly 40 miles, has been proved and mapped; and, further, conglomerate beds of a similar character, also auriferous, have been found in several outlying districts such as the Heidelberg, Klerksdorp, Venterskron, and Potchefstroom districts.

Before describing the conglomerate beds of the Rand proper, we will first give a few particulars of these outside districts.

Heidelberg District.—Under this name are included two in reality distinct areas in which the auriferous conglomerates of the Rand Formation occur. One of these is the Nigel District, with which may be included Botha's Kraal and the Town-lands of Heidelberg, and the other the detached syn-

[1] A. Schenck, *Zeitschrift der Deutschen Geologischen Gesellschaft*, 1889, p. 578.

[2] G. A. F. Molengraaff, *Neues Jahrbuch für Min., Geol., und Palæont.*, ix. Beilage-Band, 1894, p. 227.

clinal area lying to the south-east of Heidelberg, between that town and Standerton on the Natal Coach Road. In the former we have a series of sandstones and quartzites dipping at an angle of 30° to the north and underlaid by shales. Between the quartzites and the shales is a thin seam of conglomerate or banket, carrying gold in considerable quantity. This is the famous Nigel Reef. Other and thicker beds of conglomerate occur higher up in the series, outcropping further to the north. The property of the Nigel Gold Mining Company is the mynpacht or freehold mining area of the farm Varkensfontein, situated about 12 miles north-east of Heidelberg and 35 miles south-east of Johannesburg. In the north-eastern portions of this property the strike of the reef is approximately east and west, but farther east the beds soon acquire a northerly trend, and after being traceable for a short distance on Marievale, pass under the coal measures of the Karoo formation. It is possible that were it not for the presence of the Karoo formation, a connection between these conglomerates and those outcropping on the Rand might be traced. South of the Nigel property the conglomerate formation pursues a somewhat serpentine course, outcrops having been found on the farms Marais' Drift, Portje, Botha's Kraal, Houtpoort, Heidelberg Town-lands, Elandsfontein, and Schickfontein. The beds appear finally to acquire a south-westerly direction, trending towards the Vaal River, where they are again seen on the farm Strykfontein, 35 miles to the south of Johannesburg.

The other area referred to is a well-defined, oval-shaped, synclinal trough, which terminates towards the east in the property of the Heidelberg-Roodepoort Gold Mining Company. The northern margin of the syncline is on the farms Rietfontein, Rietbult, Van Kolder's Kop, and Doornhoek. The conglomerate beds outcropping on these farms strike east and west, and have a southerly dip. They are not always continuous, being in places interrupted by igneous intrusions, especially on Van Kolder's Kop and Doornhoek. The southern margin is found in Malan's Kraal, Tweefontein, Rietfontein, Daspoort, Rietvlei, Haartbeestfontein, Rooderval, and the Hex River. The length of the syncline is, east and west, about 20 miles, while the greatest breadth is about 15 miles. Both near the eastern and the western ends of the syncline, the beds are much disturbed by the intrusion and outpouring of vast quantities of igneous rock of diabasic character. At the extreme eastern end, however, a portion of the sedimentary formation remains intact, and a conglomerate reef resting on shales has been opened up on the properties of the Heidelberg-Roodepoort, and Kildare Gold Mining Companies. The dip of the beds on these properties is towards the west. With the exception of these two mines, the work done in the district has been small, and the question whether the conglomerate reefs, which undoubtedly exist, will prove generally payable, has yet to be determined. A considerable amount of money has during the last few months been put into this district; and the active prospecting which is now going on ought to definitely decide whether this is to be a great gold-producing district or not.

The Klerksdorp or *Schoonspruit District.*—The Klerksdorp Goldfields comprise several outlying patches of quartzites and conglomerates in some respects similar to the Rand Formation. The beds have been much disturbed by enormous outpourings of igneous rock (amygdaloidal lava or melaphyre). South of Klerksdorp they appear to pass under the horizontal white sandstone of the Karoo formation. The chief districts are the Wolverand, Boschrand, and Buffelsdoorn. The Wolverand District is situated about 12 miles west of Klerksdorp, and embraces the farms Rietkuil, Wolverand, and Elandslaagte.[1] The reefs form a small oval-shaped synclinal trough, of which the long axis, directed north-east and south-west, is 8 miles in length, and the short axis 1

[1] We are indebted to Mr. R. de Bonand for notes on this district.

mile. The reefs dip towards the centre of the basin at angles varying from 25° to 35°. They crop out on the northern, western, and southern sides of the basin, but are hidden on the eastern side. At the northern apex of the syncline, the reefs are being exploited by the Afrikander, on the eastern side by the Wolverand, and near the southern end by the Elandslaagte Company. On the ground of the Wolverand Company there are six reefs, the smallest of which is 2 feet wide, the largest 7. On the Elandslaagte Property, 9 feet of banket are worked.

The beds of the Boschrand Series deviate little from the horizontal, the dip rarely exceeding 15°. The reef consists of medium-sized pebbles, and is highly ferruginous where oxidised, and very pyritic below the oxidised zone. It is usually and probably correctly correlated with the Black Reef. It has been traced over a considerable extent of country, and is being worked by the Eastleigh, Ariston, and Foxdale Companies. Other companies are being formed to explore ground in this vicinity.

The Buffelsdoorn Reef, where opened, strikes north-east and south-west, dipping at an angle of about 35° to the south-east. The pay-reef is 3 to 4 feet in width, and consists mainly of a coarse-grained quartzite. On the footwall there is a layer of largish pebbles. The gold contents are not, however, confined to this pebble seam, being chiefly present in streaks in the overlying quartzite. A considerable proportion of the gold is carried by narrow seams of black bituminous or coaly matter, mineralised with arsenical pyrites.[1] This remarkable ore-body is being worked by the Buffelsdoorn Company on the farm of the same name. It has also been opened on the Government farm north of the Eleazar and on Palmietfontein. With regard to the cor-

[1] We have been shown a most interesting specimen of this substance. It is about an inch in thickness, and has a strong resemblance to a bituminous coal. On the cleaved faces streaks and films of gold are clearly visible.

relation of this deposit with the Witwatersrand beds, our knowledge is yet too imperfect to permit of this being attempted. There are nevertheless exponents of the theory that we have in the Buffelsdoorn Reef the south-westerly continuation of the Main Reef Series; while it has also been suggested that these deposits are identical with the Rietfontein Reefs, and that the Main Reef Series has still to be discovered.

The Venterskron and Free State District.—The geological structure of this district may be briefly described as a synclinal basin formed by a belt of the conglomerate formation dipping on all sides towards a common centre. The central portion of the basin is occupied by a circular boss of granite about 21 miles in diameter on which are situated the towns Verdefort and Parys in the Free State. Immediately surrounding the granite is a narrow bed of white quartzites and banded ironstone, then follows a broader belt of reddish sandstone and quartzites with associated beds of conglomerate similar to the Rand formation. Beyond this there is in places a narrower zone of volcanic lavas. The Transvaal portion of the basin lies to the north of the Vaal River, Venterskron forming its central point. The dip of the beds, which varies from 30° to 80°, is southwards. The farms in the Transvaal traversed by the reefs are the following:—Kromdraai on the west, followed by Nooitgedacht, Rooderand, Tigerfontein, Buffelskloof, Koedoesfontein, Buffelshoek, Wonderboom, and Witkopjesfontein. From Witkopjesfontein the formation crosses the Vaal River on to the farm Val Wal in the Orange Free State.

On Tigerfontein and Koedoesfontein the conglomerates attain their greatest thickness. A section across the formation on these farms would include some 10 or 15 banket beds, of widths varying from a few inches to 10 or 15 feet. Westwards and south-westwards the number of beds decreases to some extent, but the more marked thinning is to be observed eastwards towards Witkopjesfontein, where a section includes only some half-dozen conglomerate beds, mostly under 3 feet in thickness. The extent along the strike of the Transvaal portion of the formation in the Transvaal is about 29 miles. Although occasionally a promising strike has been made on these reefs, the bulk of the banket is of very low grade. Considerable work has been done on the Amazon property in this line; and other companies have recently been formed in the neighbourhood.

In the Free State the value of the reefs has not been proved to any great extent, although some prospecting is being done on several of the farms near the Vaal River.

CHAPTER III

THE AURIFEROUS CONGLOMERATES OR "BANKET"[1] BEDS OF THE RAND

THE conglomerate beds of the Witwatersrand formation may be divided for purposes of convenient description into two groups, namely :

1. Those that show conformable stratification.

2. Those that are unconformable, or of which the stratigraphic position has not been finally determined.

The first group comprises the following series of beds, arranged in descending order :

1. Elsburg Series.
2. Kimberley Series.
3. Livingstone and Bird Reef Series.
4. Main Reef Series.

In the second group we include the—

1. Rietfontein Series.
2. Botha's Series.
3. Battery Reef Series.
4. Black Reef Series.
5. Vulcan and Lanham Series.
6. Steyn Estate Series.

The members of the first group may be traversed successively (in ascending stratigraphical order) by walking from Johannesburg across the formation in a southerly direction. The outcrops of the Main Reef Series lie immediately south of the town, and are now marked by the line of trenches, open cuts, inclined and small vertical shafts, relics of the infancy of

[1] The name banket was given to these beds by the Boers on account of their resemblance to "almond rock" (for which "banket" is the Dutch name).

the industry of these fields. From the Main Reef Series to the Bird Reef Series 3000 feet of outcropping sandstones and quartzites are crossed. Between the outcrops of the Bird Reef Series and those of the Kimberley Series another 5000 feet of sandstones and quartzites intervene. Thence to the Elsburg Series there are 10,000 feet of the same rocks in which occur several unimportant conglomerate beds, sometimes designated as the Klipportje Series.

The Elsburg Series.—This series of conglomerates is named after the village of Elsburg, which was built near their outcrop. The beds are well seen in section along the Heidelberg Coach Road, and they have also been exposed in numerous cuttings on Rass's farm south of the Turffontein Estate, where they form a range of low hills. The series comprises a considerable number of beds, of which the component pebbles, consisting of white quartzites and quartz, are unusually large, averaging several inches in diameter, while some attain to even much larger dimensions. Dipping at a low angle to the south, the Elsburg beds are soon overlaid by an immense overflow of igneous rock of basaltic composition known as the Klipriversberg Amygdaloid, which forms the picturesque range extending westwards through Vierfontein to the heights of the Eagle's Nest. The gold contents of the Elsburg beds is small, and although in the early days many attempts were made to work them, no great measure of success was attained. The greatest amount of work was done on the property of the Aurum Company, situated on the farm Elandsfontein No. 2, and on the property known as the "Golden Quarry Homestead," or De Pass's, on Rass's farm.

It is instructive to compare with the Elsburg Series the conglomerates occurring north of the Main Reef Series in the hills forming the south side of the Doornfontein valley north of Jeppe's Town. Walcot Gibson has pointed out the resemblance of these conglomerates to those of the Elsburg

Series. He throws out the suggestion of their identity, and shows how the Doornfontein beds might have been brought into their present position by faulting (see Fig. 1, which is taken from his paper). According to this view there are two alternatives, either the beds have been faulted down or they have been brought up from below. We endorse the first alternative, as the second involves the inversion of the whole sequence of beds composing the Rand formation.

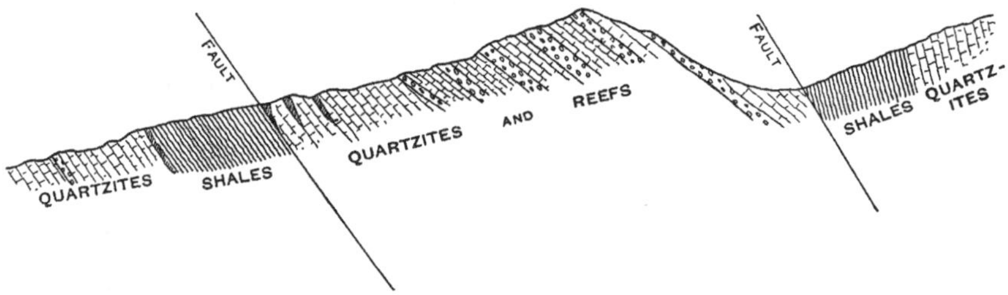

FIG. 1.—Section across the hills north of Jeppe's Town. *After Walcot Gibson.*

The Kimberley Series.—This belt of conglomerates lies some 8000 feet south of the outcrop of the Main Reef Series. It comprises a number of banket beds, some of which are of considerable size. Several of them have received distinctive names, suggested in some cases by individual peculiarities; amongst them are the Yellow Reef, Red Reef, Sunday Reef, Free State Reef, and Kimberley Reef. The component pebbles are of a fairly large size, and consist of white and gray quartzite, banded and indurated shale, and more rarely of quartz. The matrix in which the pebbles are embedded is usually a gritty or sandy material. In some places small beds of shale are associated with the conglomerate beds.

In the early days the Kimberley beds were opened at various points along the line, and a number of gold-mining companies were formed to work them.[1] The ore bodies,

[1] The Gordon Estate, Great Britain, Marie Louise, Red Reef, Mitchell, Klipriversberg Estate, Ziervogel, and Leewpoort Companies are examples.

STANHOPE & STANHOPE-GELDENHUIS DEEP.

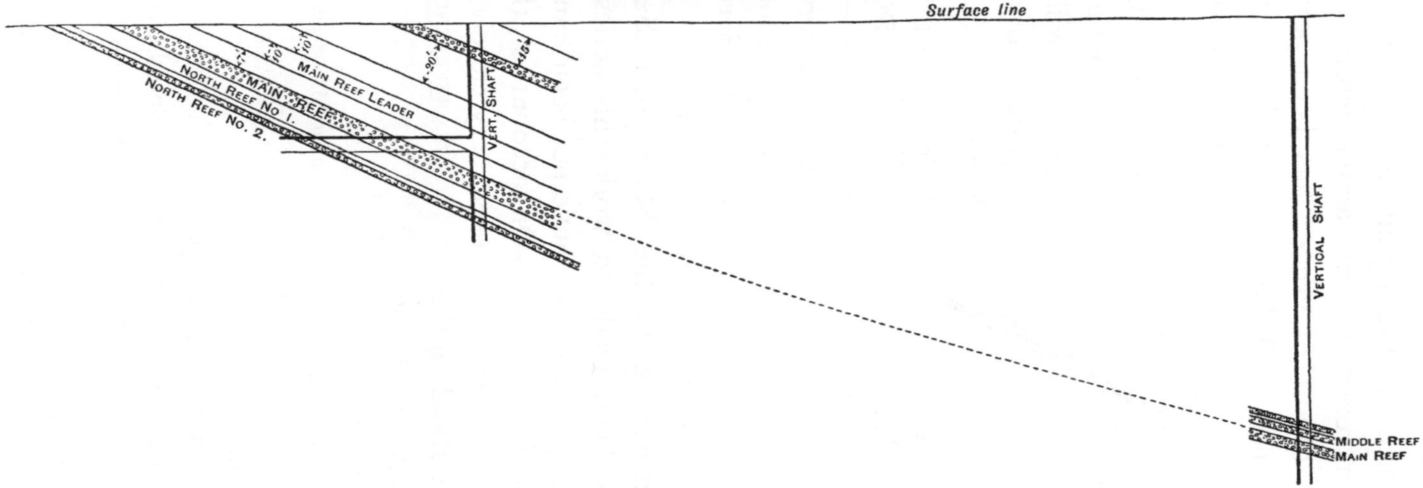

FIG. 2.

however, proved of too low a grade to pay under the economic conditions then existing, and active work was soon suspended. The ground, however, has been persistently held under claim licenses, and there is a belief current that the day is rapidly approaching when reductions in working expenses, and improvements in the methods of gold extraction, will enable considerable stretches of the Kimberley Series to be worked at a profit.

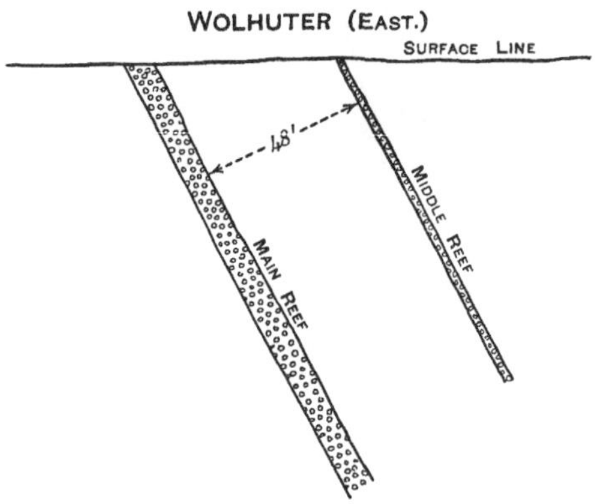

FIG. 3.

The Livingstone and Bird Reef Series. — This series consists of seven or eight beds of conglomerates composed of smaller pebbles than the Kimberley Series, and lying between that and the Main Reef Series. The pebbles are chiefly quartz, as in the Main Reef Series. Some attempts have been made to work this series on Paardekraal;[1] but the gold contents have not been found payable.

The Main Reef Series.—On account of their valuable gold contents these reefs have been very closely studied from one end of the Rand to the other, and they are being worked by a series of gold-mining companies extending in close succession from near Boksburg, 13 miles east of Johannesburg, to Krugersdorp, 18 miles west of the same town. If we also include those companies working on the east and west extensions beyond Boksburg and Krugersdorp,

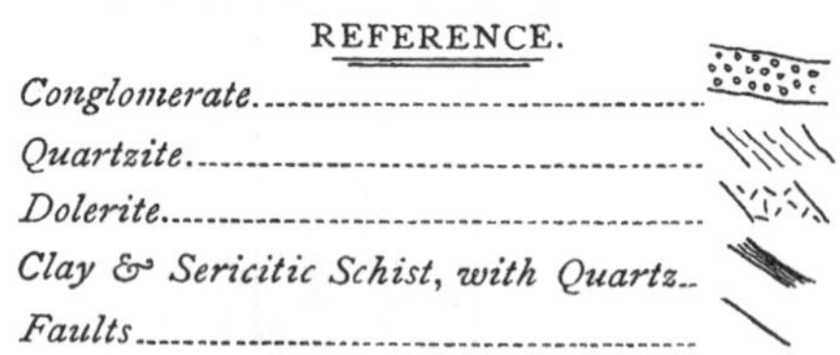

FIG. 4.—Geological Formation of New Primrose Mine.

[1] The Citizen, Western Chimes, United Chimes, Eva, Exchequer, Ida, Johannesburg Roodepoort, Paardekraal, and Cordova were companies formed to work the Bird Reef Series.

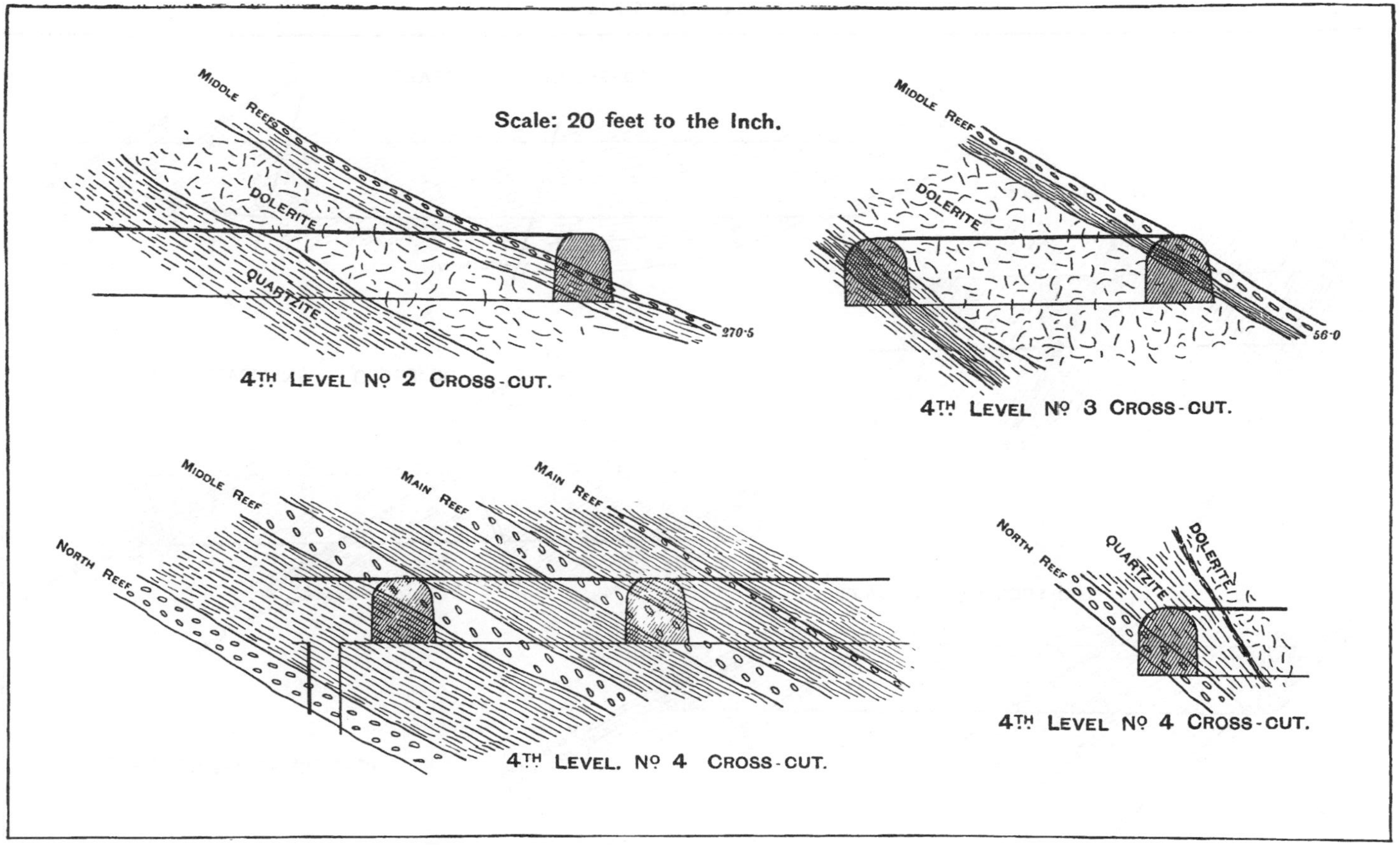

FIG. 5.—Sections of the Reefs worked in the New Primrose Mine.

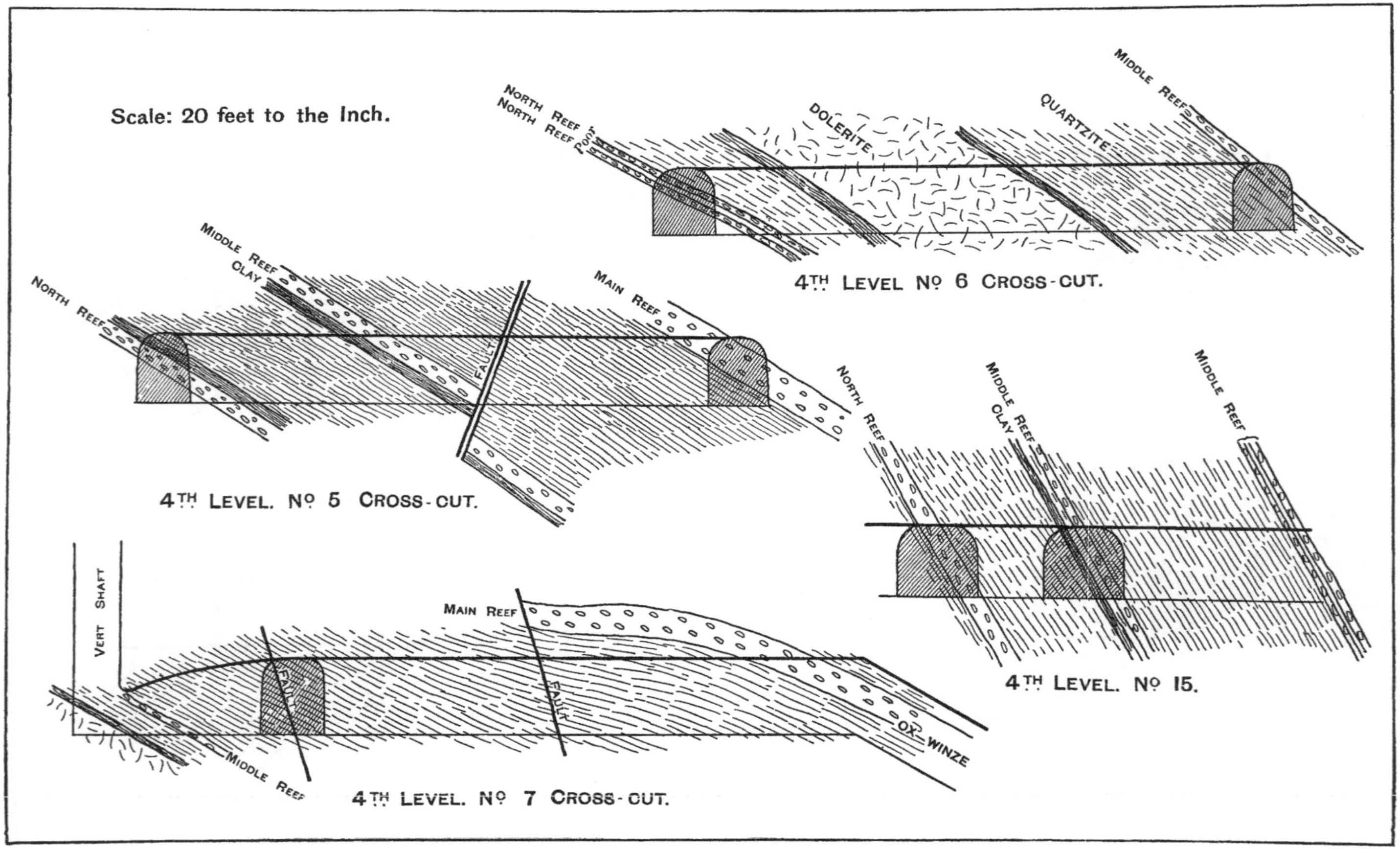
Scale: 20 feet to the Inch.
North Reef
North Reef
Dolerite
Quartzite
Middle Reef
4TH Level No 6 Cross-cut.
Middle Reef
Clay
North Reef
Main Reef
Fault
4TH Level. No 5 Cross-cut.
North Reef
Middle Reef
Clay
Middle Reef
4TH Level. No 15.
Vert Shaft
Main Reef
Fault
Fault
Ox-Winze
Middle Reef
4TH Level. No 7 Cross-cut.

Fig. 6.—New Primrose.

the whole distance covered by companies in active operation amounts to about 46 miles.

To enable the reader to obtain a clear insight into the extent and positions of the workings on the Main Reef Series, we give the following list of outcrop companies arranged in serial order east and west of Johannesburg, the name of the farm on which each property lies being given as well as the length of the outcrop on each farm.

PROPERTIES IN THE IMMEDIATE NEIGHBOURHOOD OF JOHANNESBURG AND EAST OF JOHANNESBURG

Name of Company.	Name of Farm.	Length of Outcrop.
Mint. Johannesburg Pioneer. Robinson. Worcester. Ferreira. Wemmer. Salisbury. Treasury (part). Jubilee. City and Suburban (part).	Turffontein.	2.2 miles.
City and Suburban (part). Meyer and Charlton. Wolhuter. Spes Bona. George Goch. Metropolitan. Henry Nourse. Ruby. New Heriot. Jumpers. Treasury (part).	Doornfontein.	4.0 miles.
Geldenhuis Estate. Geldenhuis Main Reef. Stanhope. Simmer and Jack. New Primrose (and Moss Rose). May Consolidated (part).	Elandsfontein.	2.6 miles.

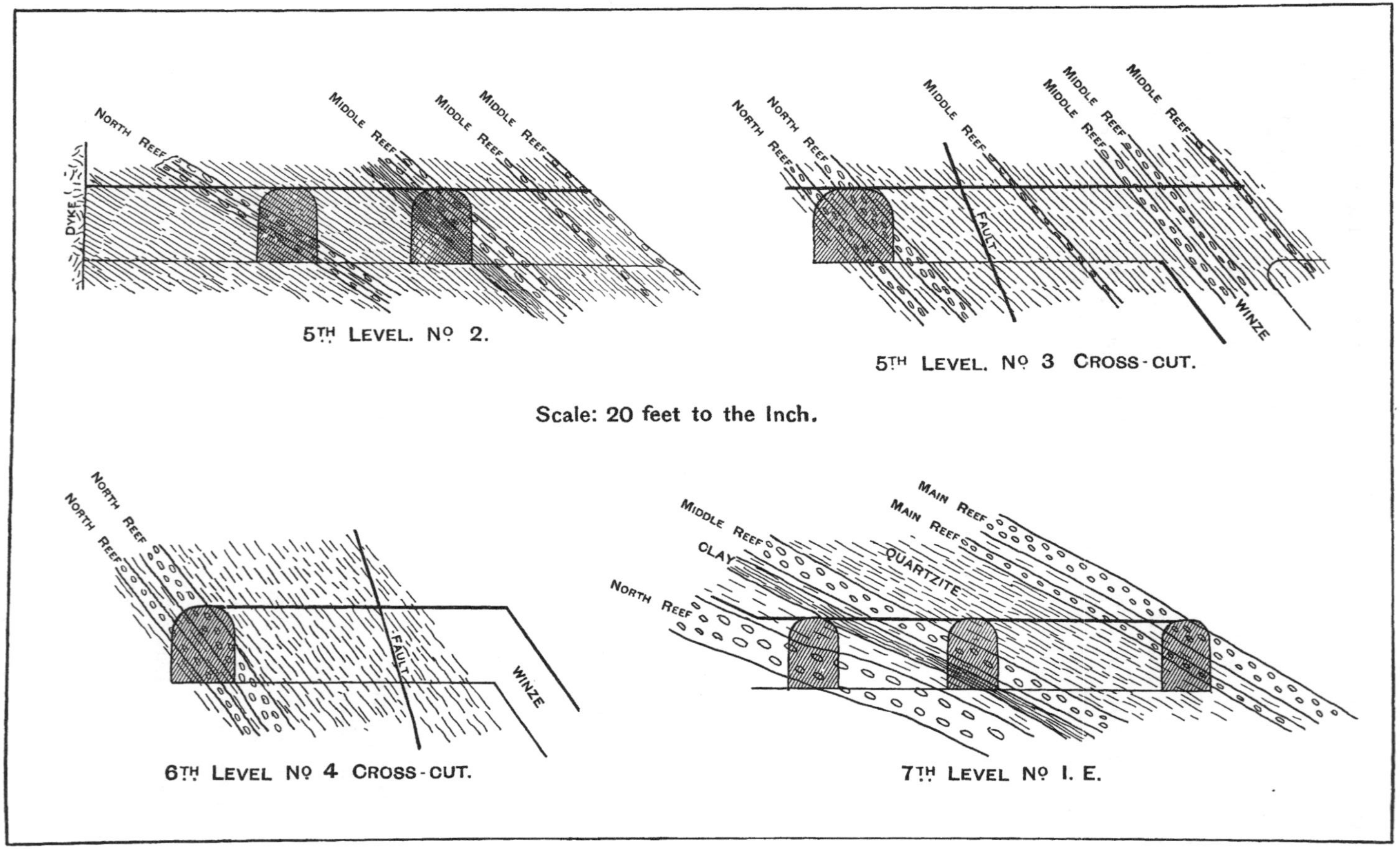

North Reef
Middle Reef
Middle Reef
Middle Reef
Dyke
5TH LEVEL. No. 2.
North Reef
North Reef
Middle Reef
Fault
Middle Reef
Middle Reef
Middle Reef
Winze
5TH LEVEL. No. 3 CROSS-CUT.
Scale: 20 feet to the Inch.
North Reef
North Reef
Fault
Winze
6TH LEVEL No. 4 CROSS-CUT.
Middle Reef
Clay
North Reef
Quartzite
Main Reef
Main Reef
7TH LEVEL No. I. E.

FIG. 7.—New Primrose.

Name of Company.	Name of Farm.	Length of Outcrop.
May Consolidated (part).		
Glencairn.		
Knight's Tribute.		
Witwatersrand (Knight's).		
Balmoral.		
Gardner.	Driefontein.	5·5 miles.
Ginsberg.		
Driefontein.		
St. Angelo.		
Comet.		
Agnes Munro.		
Cason Block.		
Cinderella.	Vogelfontein.	1.2 miles.
Blue Sky.		

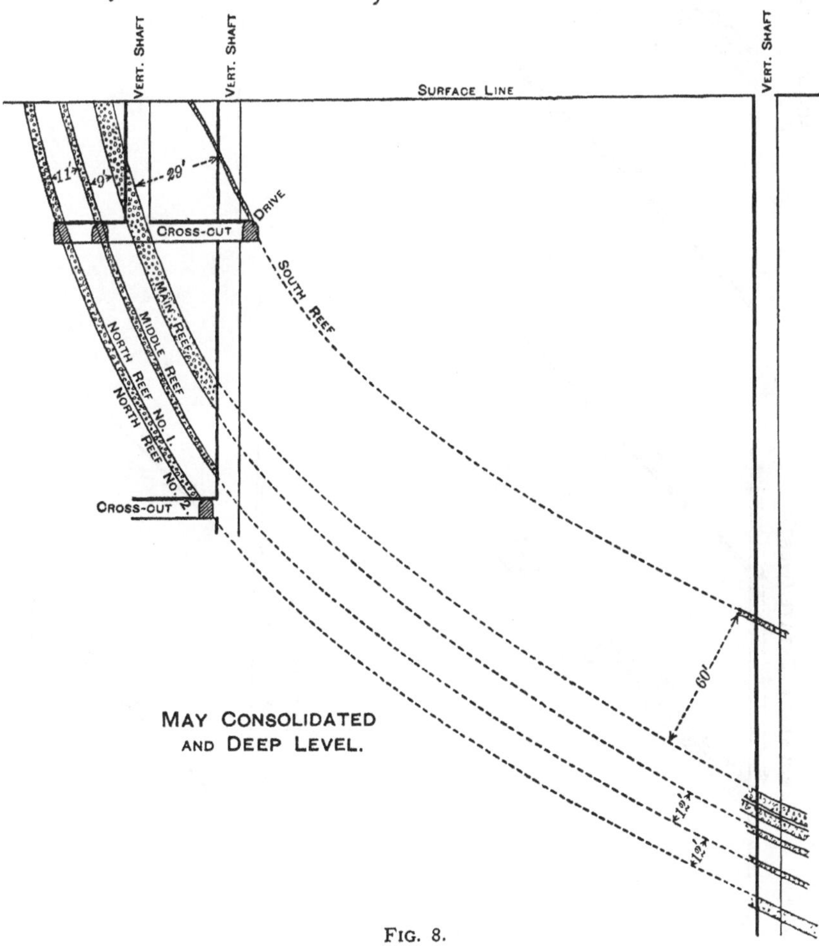

Fig. 8.

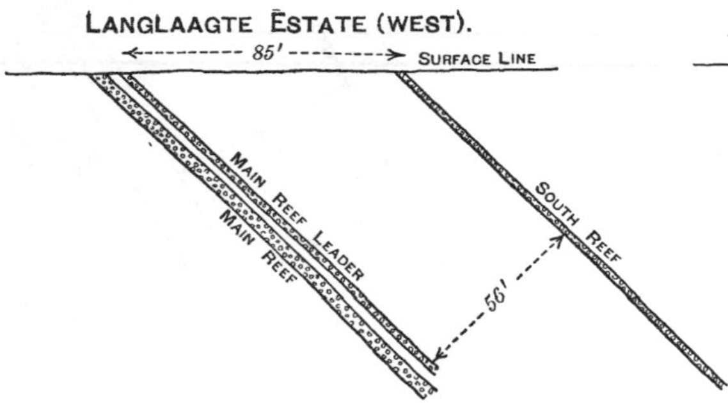

FIG. 9.

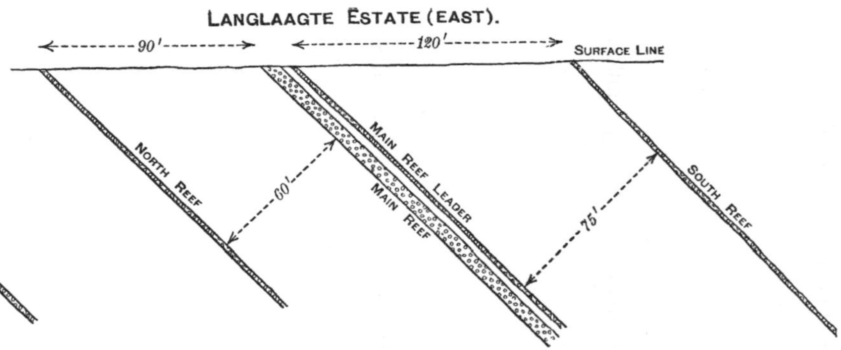

FIG. 10.

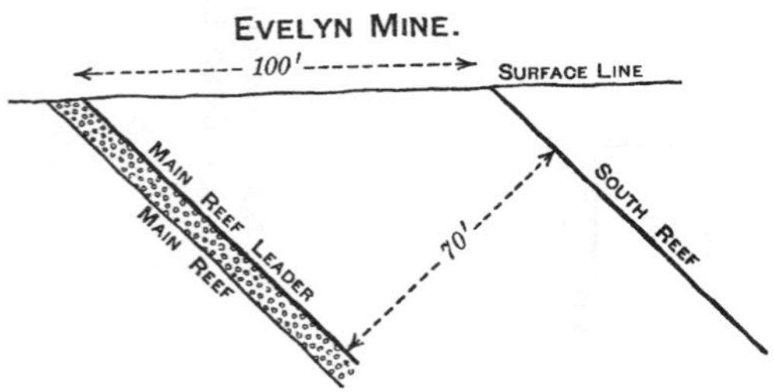

FIG. 11.

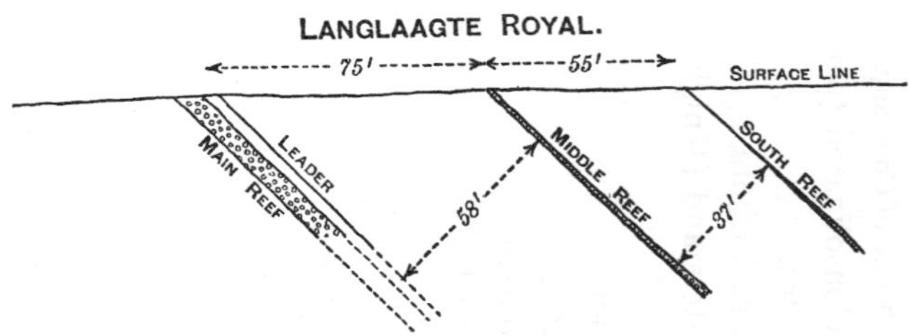

FIG. 12.

Outcrop Properties on the Eastern Extension and Chimes District

Name of Company.	Name of Farm.	Length of Outcrop.
Kleinfontein.	Kleinfontein.	4.6 miles.
Van Ryn.	Vlakfontein.	
Chimes.	Benoni.	
Modderfontein.	Modderfontein.	

Outcrop Properties West of Johannesburg

Name of Company.	Name of Farm.	Length of Outcrop.
Crown Reef. Langlaagte Estate. Langlaagte Royal. Paarl Central. United Langlaagte. Langlaagte Block B. New Crœsus. Langlaagte Star.	Langlaagte.	3.8 miles.
Tharsis, Angle Tharsis, Nabob (Now consolidated Angle Tharsis.) Main Reef New Unified. Aurora. Aurora, West, United.	Paadekraal.	2.7 miles.
Odessa. Bantjes. Vogelstruisfontein.	Vogelstruisfontein.	2.6 miles.
Kimberley Roodepoort. Durban Roodepoort. Roodepoort, United (including the Evelyn). Princess. Banket (part).	Roodepoort.	3.4 miles.
Banket (part). Bohemian. Gipsy.	Witportje.	0.9 mile.

Outcrop Properties on Western Extension.

Name of Company.	Name of Farm.	Length of Outcrop.
Teutonia. Champ d'Or.	Witportje.	1.1 miles.

WEMMER.

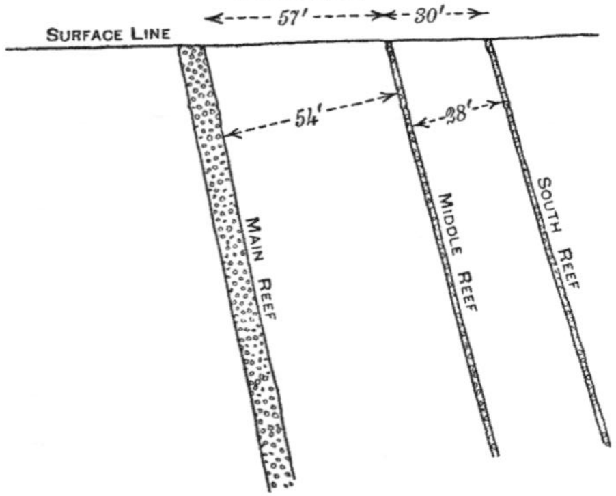

Fig. 13.

GLENCAIRN

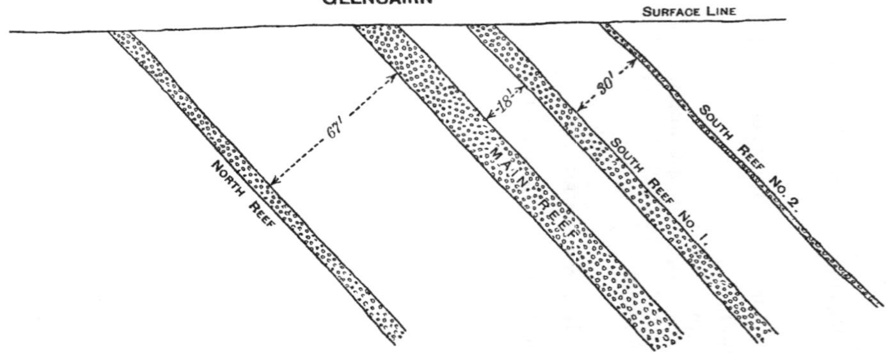

Fig. 14.

FERREIRA.

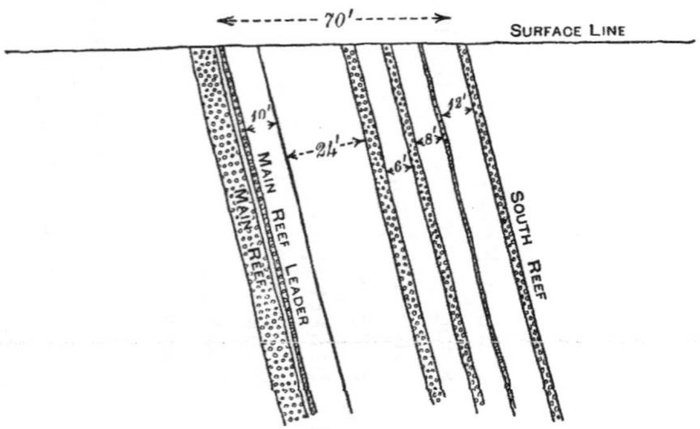

Fig. 15.

DURBAN ROODEPOORT.
(No. 2 Block.)

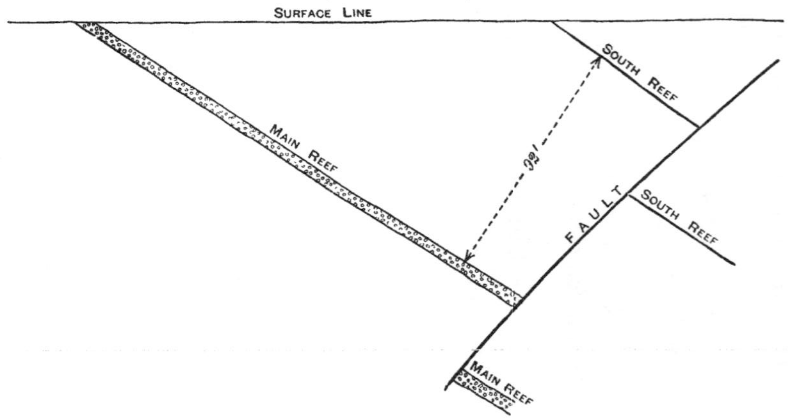

Fig. 16.

Name of Company.	Name of Farm.	Length of Outcrop.
Windsor (late Britannic) Luipaardsvlei Estate. York (late Emma). George and May (late Bothas) (part).	Luipaardsvlei.	4.4 miles.
George and May (late Botha's) (part). Randfontein Estate (part).	Waterval.	1.5 miles.
Randfontein Estate (part). Robinson Randfontein. North Randfontein.	Uitvalfontein.	3.3 miles.
Porges Randfontein.	Randfontein.	2 miles.

SUMMARY

East of Johannesburg (to Boksburg)	15.5	miles	of outcrop.
West of Johannesburg (to Witportje)	13.4	"	"
Eastern Extension (Modderfontein, etc.)	4.6	"	"
Western Extension (Luipaardsvlei, Randfontein, etc.)	12.3	"	"
TOTAL	45.8		

That the outcrop of the Main Reef Series extends throughout the length of the properties enumerated above, has been clearly demonstrated. For the greater part of the distance, the beds have been traced continuously from property to property; and where this is the case, there can be no doubt as to correct identification. Where, on the other hand, considerable breaks occur, we have of course to be guided by other evidence in picking up the series. The best testimony to a correct correlation is afforded when the identity of the overlying and underlying formations, and a similarity in the character and gold-contents of the beds comprising the series, can be demonstrated. It is true that the series present considerable variations when examined in different parts of the Rand; but this is exactly what should be expected from the nature and mode of origin of the beds. Indeed it is sufficiently remarkable that beds usually

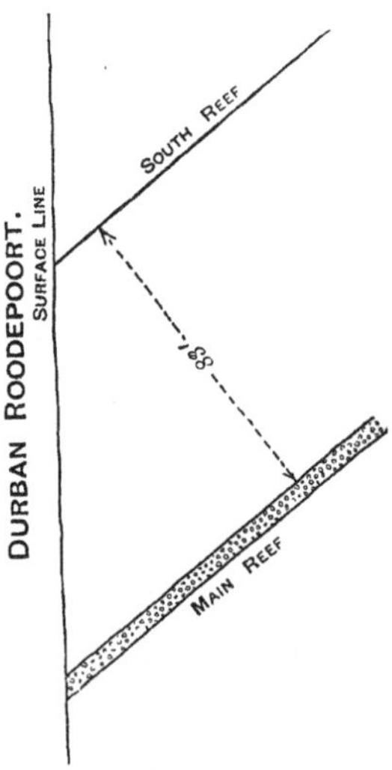
DURBAN ROODEPOORT.
SURFACE LINE
SOUTH REEF
83'
MAIN REEF

FIG. 17.

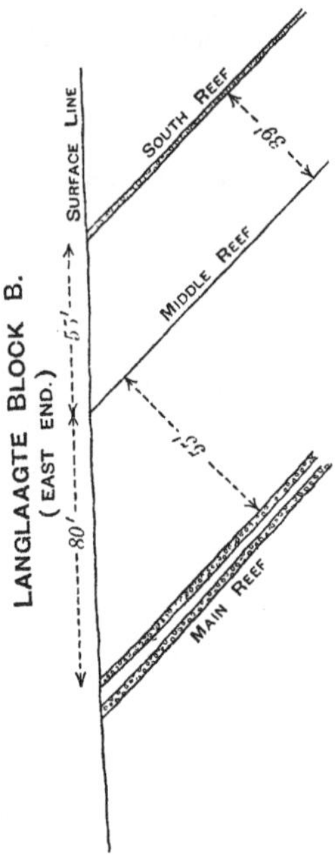
LANGLAAGTE BLOCK B.
(EAST END.)
SURFACE LINE
SOUTH REEF
39'
MIDDLE REEF
80'
57'
53'
MAIN REEF

FIG. 18.

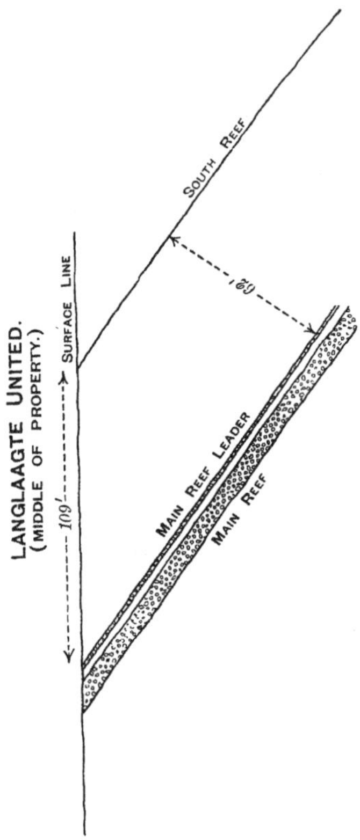
LANGLAAGTE UNITED.
(MIDDLE OF PROPERTY.)
SURFACE LINE
SOUTH REEF
62'
109'
MAIN REEF LEADER
MAIN REEF

FIG. 19.

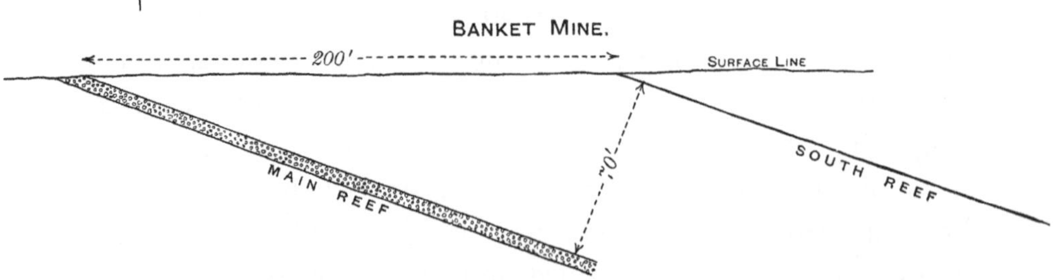
BANKET MINE.
200'
SURFACE LINE
SOUTH REEF
70'
MAIN REEF

FIG. 20.

so erratic and irregular as conglomerates should vary so little, and that thin beds, sometimes consisting of a single layer of pebbles, should remain persistent over such large areas.

In appearance the banket beds of the Main Reef Series have the usual characteristics of conglomerates, *i.e.* they are composed of round or ellipsoidal pebbles, which present every sign of having been water-rolled and worn smooth by attrition. The pebbles consist mainly of white or smoky quartz, and lie imbedded in a sandy or quartzitic matrix. In size they vary from a pea to a hen's egg, but as a rule the pebbles of any one bed are fairly uniform over considerable distances.

The changes that are observed when sections of the Main Reef Series, taken at points far removed from one another, are compared, consist chiefly in the number of the beds and in the amount of sandstone or quartzite lying between them. The angle at which the beds are lying naturally affects the apparent interval between them, measured on the level. Comparisons, therefore, to be of any service, can only be instituted when the beds are measured at right angles to the dip. The greatest constancy in this respect is found in the section of country lying between the Wolhuter on the east and the Banket Company on the west. Eastwards from the Wolhuter changes become more marked, and it becomes increasingly difficult to prove the identity of any particular reef on different properties; while west of the Banket the ground is broken up by faults, and there is considerable divergence of opinion as to what is to be considered as the true extension of the Main Reef Series.

In its most typical development, *i.e.* over the area lying between the Wolhuter and the Banket, the Main Reef Series consists of three more or less payable reefs. The lowest or most northern of these is a big body (in places as much as 12 feet thick) of low grade banket. This is termed the

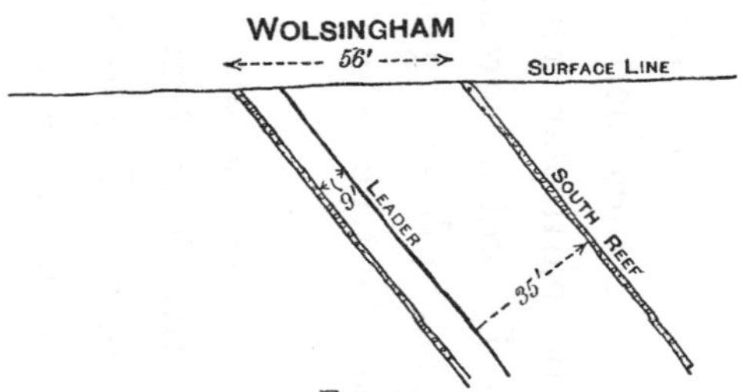

FIG. 21.

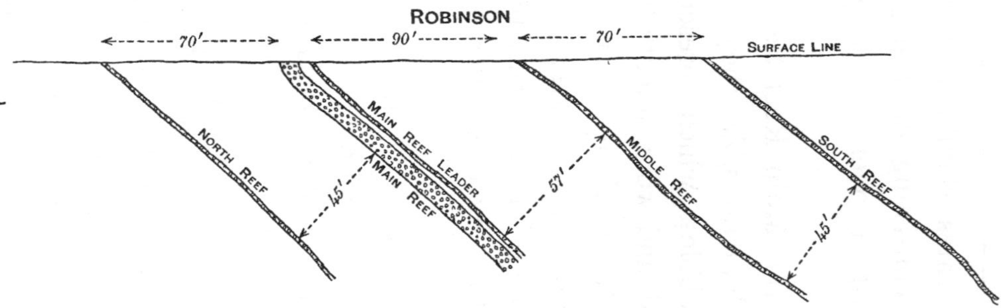

FIG. 23.

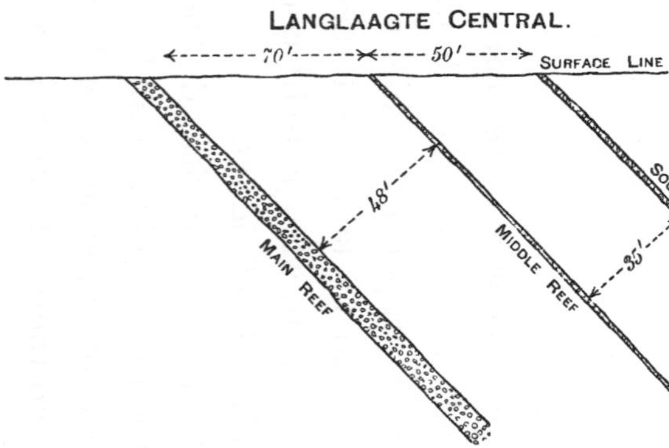

FIG. 22.

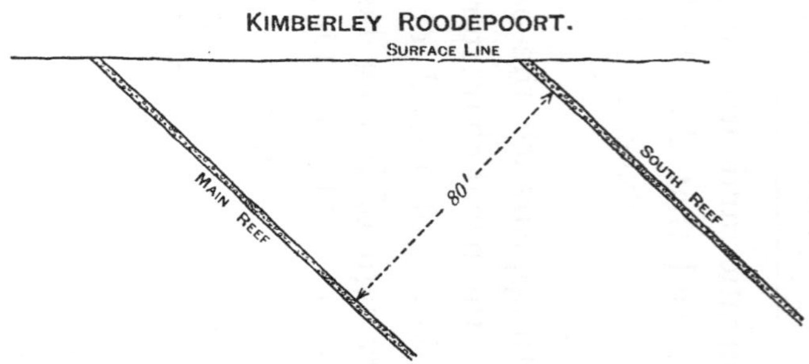

FIG. 24.

Main Reef, and has given its name to the whole series. Lying above the Main Reef (or to the south of it), and separated from it by a varying amount of sandstone, is a much smaller body of payable banket, known as the Main Reef Leader.[1] This ore-body averages about 15 inches in width, ranging from a few inches up to 3 feet. To the south of the Main Reef Leader is the South Reef, a body of banket varying from a few inches to as much as 6 feet in thickness. It is generally split up by sandstone partings into two and sometimes even three leaders, of which the footwall or "South Leader" is the most valuable. The sandstone separating the Main Reef Leader from the Main Reef is in some places as much as 6 or 7 feet; while in some mines it thins out to nothing, and the Main Reef Leader is then found in close juxtaposition to the Main Reef. In some few cases, *e.g.* at the Johannesburg Pioneer, the parting between the two reefs is one of clay instead of sandstone. The distance between the South Reef and the Main Reef Leader is much greater, ranging from 35 to 100 feet, measurements being taken at right angles to the slope of the reefs.[2]

Of these three reefs, two are in most cases payable, while the third (the Main Reef) is only payable in certain mines

[1] The word leader is a local term for a small reef, borrowed from vein mining.

[2] The following distances have been observed between the two reefs, in the mines mentioned, measurement being at right angles to the dip:—

	Feet.		Feet.
Wolhuter	50	Langlaagte Estate	56-76
City and Suburban	65	Crœsus	90
Salisbury	54	Star	105
Wemmer	82	Main Reef	100
Ferreira	65	Unified	100
Jubilee	52	Aurora	110
Wolsingham	35	Vogelstruis	80
Robinson	50	Banket	70

The sections of the Main Reef Series shown in Figs. 2-42 are actual sections taken in the different mines, and will serve to illustrate the different conditions as to width, distance apart, etc., obtaining at different parts.

SALISBURY.

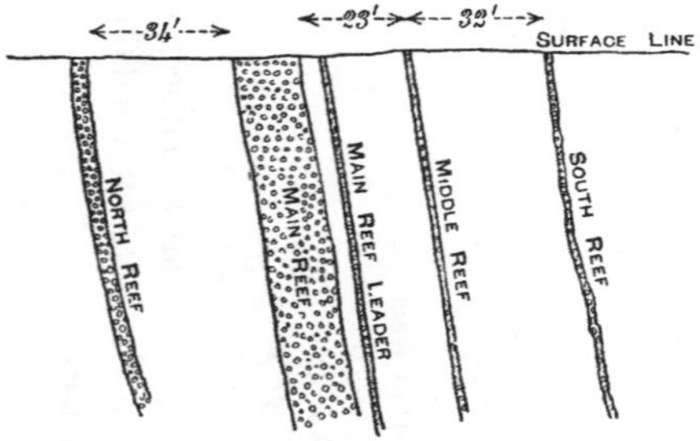

FIG. 25.

AURORA WEST MINE.

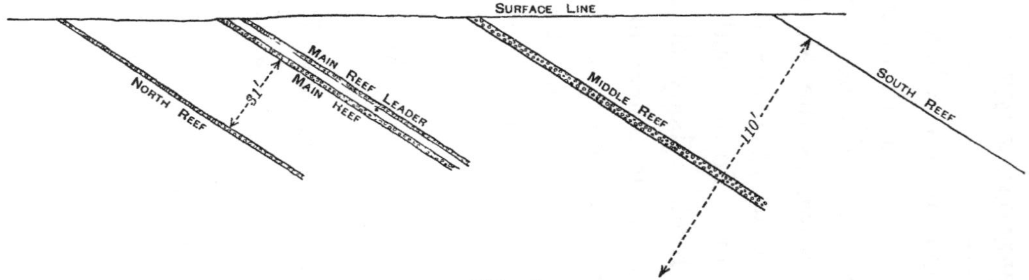

FIG. 26.

CITY AND SUBURBAN

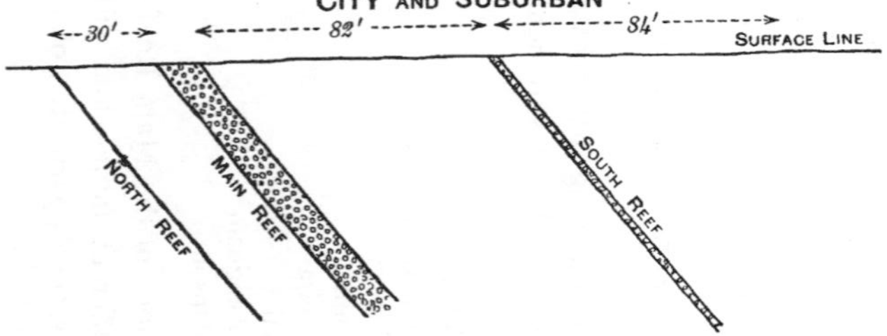

FIG. 27.

AURORA MINE.

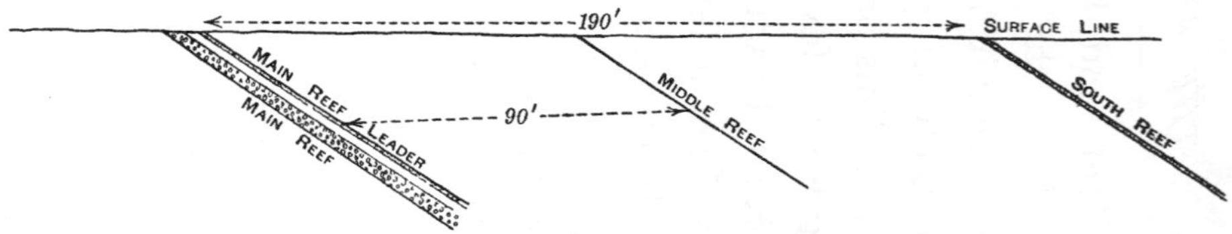

FIG. 28.

and when worked on a sufficiently large scale. Besides these ore-bodies there are frequently present other non-payable beds of conglomerate. Thus in some of the mines belonging to the section under consideration there is a "North Reef," lying to the north of the Main Reef; while in others there is a reef (Middle Reef) lying between the Main Reef Leader and the South Reef. In no case, however, have these beds been found payable.

With regard to the distinguishing characteristics of the ore-bodies in this part of the Rand, we note in the Main Reef its great size, its small gold-contents, its interbedded seams of sandstone, and the uniform size of its pebbles, which consist of white quartz and are of medium size. The Main Reef Leader, on the other hand, is of small size, but comparatively rich. The pebbles consist of white and smoky quartz. Usually there is a layer of largish pebbles on the foot-wall. Both foot and hanging walls are well defined. The South Reef in most of these mines is separable into a foot-wall seam, which carries the major portion of the gold, and a larger body of banket of lower grade. In some cases there are three leaders, two of which are poor, while the third (the foot-wall leader) is payable. The South Reef Leader seldom exceeds 8 inches in thickness, and is frequently represented by a mere layer of pebbles, or even by a thin ferruginous seam, especially in the mines lying between the Crœsus and the Banket. The pebbles of the South Reef are comparatively small, the largest being in the leader. There is a large proportion of smoky quartz among them. The sandstone parting between the South Leader and the upper seams of the South Reef varies from 1 inch to as much as 6 feet, the average being about 20 inches.

Turning now to the East Rand, we find different conditions prevailing. The South Reef becomes of less and less importance as we proceed east of Johannesburg; although still profitably worked as far as the Heriot. Farther east as

STAR.

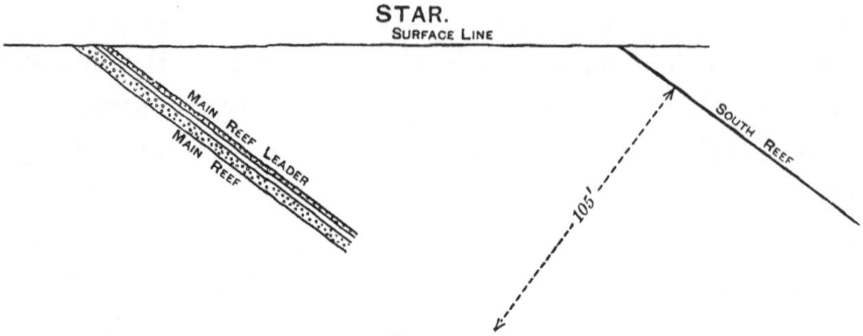

FIG. 29.

UNIFIED.

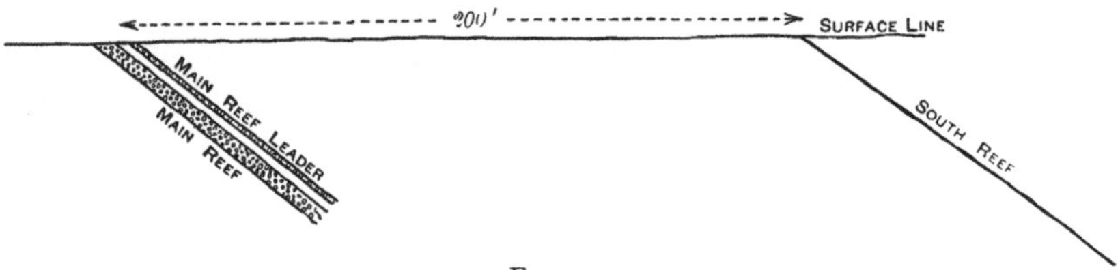

FIG. 30.

LANGLAAGTE WESTERN.

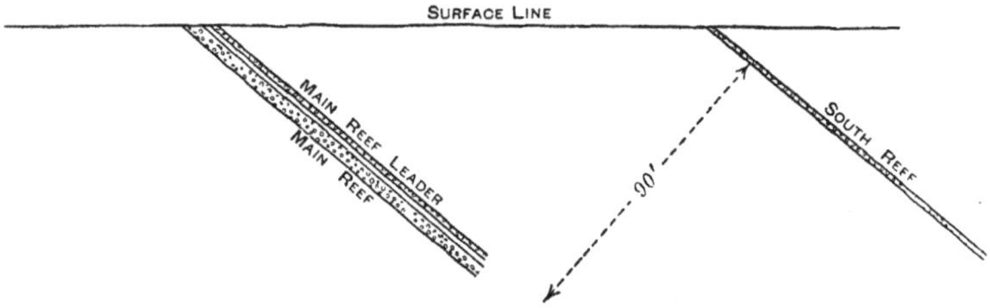

FIG. 31.

JUBILEE

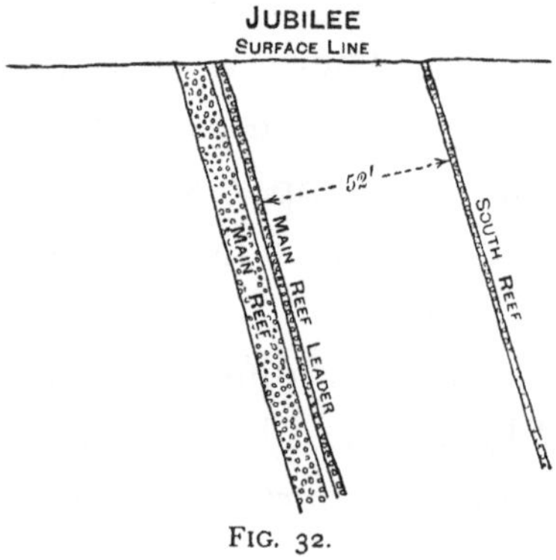

FIG. 32.

PRINCESS MINE (EAST END OF MYNPACHT).

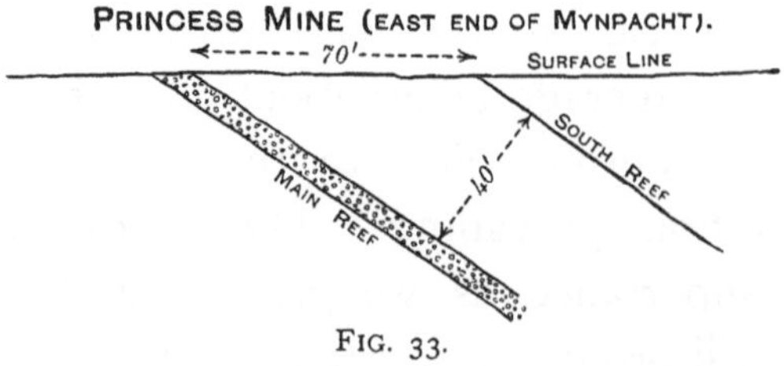

FIG. 33.

far as Knight's it is not mined at all. Apart from the South Reef we find, generally speaking, three reefs in close proximity; these are termed respectively the North, Middle, and Main, or in some mines, the North, Main, and Middle. It is difficult to correlate the reefs occurring in different properties, and attempts at uniform nomenclature, implying such correlation, are apt to be misleading. There are, however, certain distinctive features that may be traced for considerable distances. Thus a seam of clay and sparry quartz in the foot-wall of the Middle Reef enables it to be easily recognised throughout the properties of the Simmer and Jack and Primrose Companies. While through the ground of the May Consolidated, Glencairn, Knight's Tribute, and Witwatersrand Companies, a layer of schistose or slaty material underlies the same reef, which is known in these properties as the Slate Leader. Another distinctive feature, in the same properties, is the peculiar conglomerate with sparsely scattered pebbles, lying between the Slate Leader and the North Reef, and known locally as the Bastard Reef.

It will perhaps conduce to a clearer insight into the nature of the reefs worked in this part of the Rand if we describe a few typical sections through some of the more important mines east of Johannesburg.

Beginning with the George Goch and Metropolitan mines, we find, as the lowest member of the series—the North Reef, a bed of conglomerate about 4 feet thick. Separated from this by a variable amount of sandstone (2 to 12 feet) is the Main Reef, a bed some 3 to 7 feet wide. Between this and the Middle Reef is a layer of 3 or 4 feet of schistose or slaty matter with quartz veins. The Middle Reef varies from 1 to 3 feet in thickness, and is separated from the South Reef by 20 to 30 feet of sandstone. The South Reef is composed of two or three bands of conglomerate, a few feet apart, of which the foot-wall leader carries the gold. Below the 200-foot level the North Reef disappears

GELDENHUIS ESTATE.

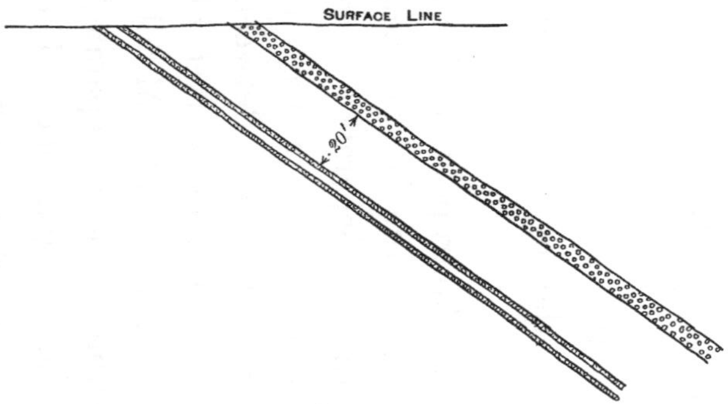

FIG. 34.

WOLHUTER (WEST.)

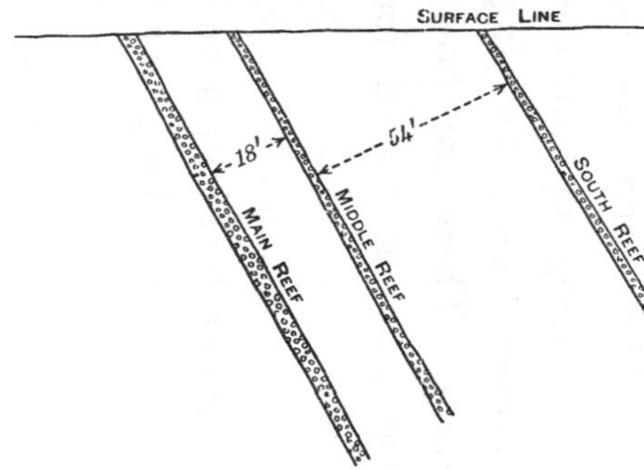

FIG. 35.

MOSS ROSE.
(NOW PART OF THE PRIMROSE.)

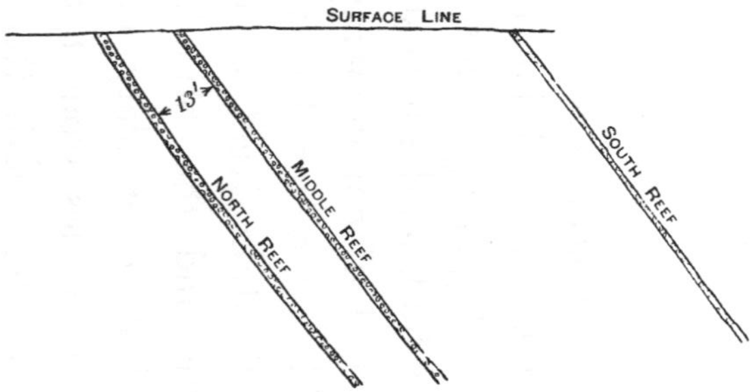

FIG. 36.

HENRY NOURSE. (WEST END.)

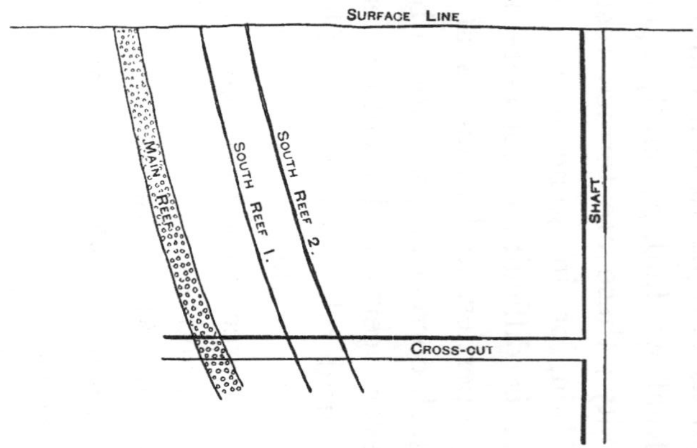

FIG. 37.

altogether: in explanation of this remarkable fact we have the choice of two theories—either it thins out, or it is an overlapped portion of the Main Reef itself.

In the properties of the Henry Nourse and Heriot Companies the series spreads out into a considerable number of beds. In the bore-hole put down by the Henry Nourse Deep Level Company, nine beds of banket were encountered within a distance of 270 feet, or 188 feet if allowance be made for the dip of the beds. The following section is taken from the Third Annual Report of the New Heriot Company. It represents the beds passed through in driving a cross-cut 50 feet south of the main shaft. The dip of the beds is about 60°. The first reef was struck at 14 feet from the foot-wall of the shaft.

Reef.	Sandstone parting.	Width.	Assay value.
No. 1	...	24 inches.	5·4
	2 inches	...	...
,, 2	...	10 ,,	14.5
	16 ,,	...	...
,, 3	...	54 ,,	27.6
	34 ,,	...	...
,, 4	...	6 ,,	4·2
	34 ,,	...	...
,, 5	...	12 ,,	18.0
	32 ,,	...	...
,, 6	...	7 ,,	4·7
	7 ,,	...	...
,, 7	...	18 ,,	2.0
	42 ,,	...	...
,, 8	...	12 ,,	3·6
	30 ,,	...	...
,, 9	...	48 ,,	8.2
	42 ,,	..	...
,, 10	...	24 ,,	3·5

In the Geldenhuis Estate the South Reef lies some 100 to 150 feet south of the uppermost reef of the Main Reef Series, and is not worked. The Main Reef Series consists practically of three bodies; the uppermost or "Main Reef" is a big body some 6 feet wide, but of too low a grade to repay working. About 40 feet to the north of this reef

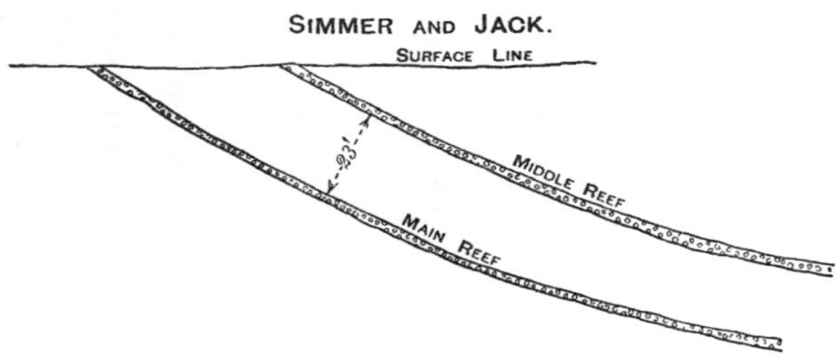
SIMMER AND JACK.
SURFACE LINE
MIDDLE REEF
MAIN REEF

FIG. 38.

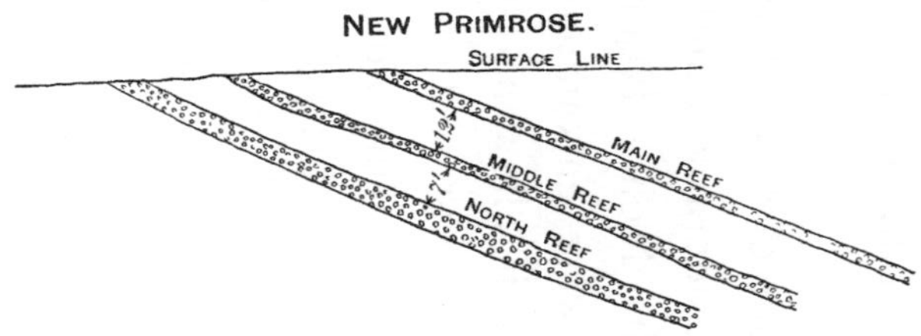
NEW PRIMROSE.
SURFACE LINE
MAIN REEF
MIDDLE REEF
NORTH REEF

FIG. 39.

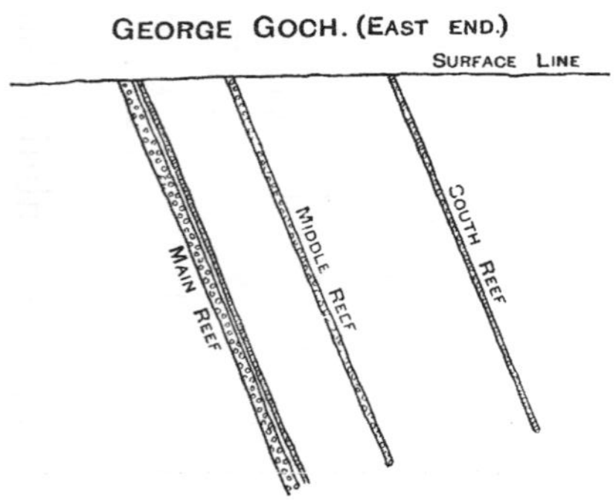
GEORGE GOCH. (EAST END.)
SURFACE LINE
MAIN REEF
MIDDLE REEF
SOUTH REEF

FIG. 40.

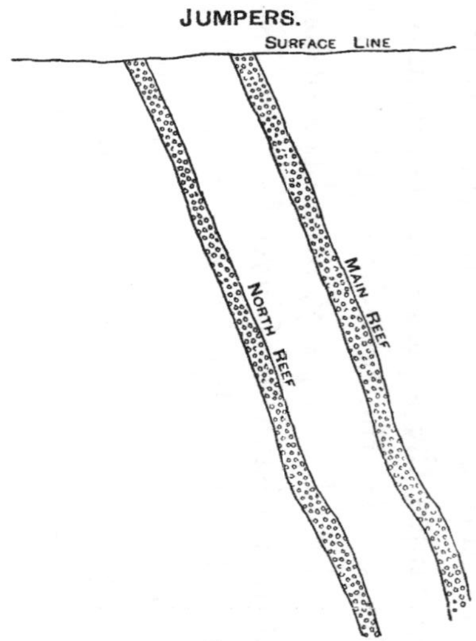
JUMPERS.
SURFACE LINE
NORTH REEF
MAIN REEF

FIG. 41.

lie the payable ore-bodies, which are taken out in one big stope. They consist of two main bodies. The uppermost of these (Middle Reef) is distinguished by a slaty parting on the foot-wall. It consists of some 15 to 20 inches of pay ore. In the western part of the mine it is split up into two small leaders separated by a seam of sandstone. Of these the foot-wall leader, although small, is very rich.

The bottom reef (North Reef) is separated from the Middle Reef by a bed of sandstone ranging in thickness from a few inches to 6 or 8 feet. This reef is also divided into two bodies, the upper of which is about 3 feet wide, the lower 16 to 20 inches in thickness. Both are of payable grade.

At the Simmer and Jack there are also three reefs: the North, the Middle, and the Main. Of these the North and the Main have hitherto yielded the best results. The distance between them varies in different parts of the mine, but all of them are generally embraced within a width of 15 to 20 feet. The size of the ore-bodies is also subject to considerable variation, and the reefs are often split by seams of sandstone into several leaders, one of which, as a rule, carries the major portion of the gold-contents. Of the three reefs the Middle is the most easily recognised on account of its carrying on its foot-wall a seam of soft clay, with which vein-quartz is associated. The South Reef lies 300 feet to the south, and although it carries some gold, has up to the present not been worked.

In the Primrose the conditions are much the same, the Main Reef Series consisting of three bodies, North, Middle, and Main; while the South Reef lies about 130 feet to the south. The Middle Reef has the same characteristic foot-wall as in the Simmer. A remarkable peculiarity of the Primrose reef, however, is the presence of a sheet of igneous rock interbedded between the North Reef and the Middle Reef, and in places squeezing out the former without affecting the latter.

In the May Consolidated, Glencairn, and Knight's Tribute, a general section shows a fair-sized body of banket forming the bottom bed; this is known as the North Reef. Two to ten feet above this, and separated from it by a peculiar rock with sparsely-scattered pebbles (the "Bastard Reef"), lies the Middle Reef, here called the Slate Leader on account of a seam of slaty matter underlying it. Some 5 or 15 feet above the Slate Leader lies a large body

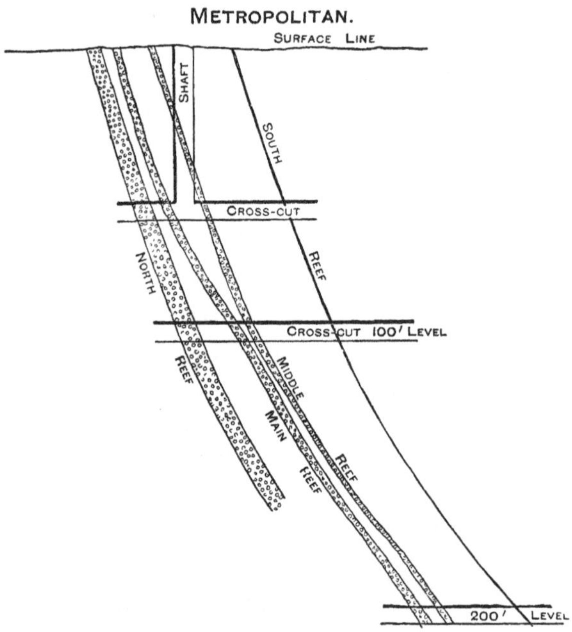

FIG. 42.

of banket known as the Main Reef. The North Reef and the Slate Leader constitute the payable ore-bodies in these mines.

In the property of the Witwatersrand Company the North Reef has contracted to a small leader, while the Slate Leader lies close to the Main Reef. The complications produced by faulting in this property will be referred to later on.

Coming now to the East Rand Proprietary Companies, we find only one reef worked in the properties of the Comet

Photographed by Davies Bros.

PLATE I.—HEADGEAR, MAY CONSOLIDATED MINE.

and Blue Sky, although it is generally separated into two bodies or leaders by a small parting of sandstone. In the Blue Sky the foot-wall of the reef rests immediately on shales. This is apparently due to a gradual thinning out of the quartzites that elsewhere lie between the Main Reef or North Reef and the first band of shales. The South Reef is also represented in this section of the Rand. It lies some 600 or 700 feet south of the North Reef, and where tested appears to be a highly payable body of banket. It is being worked at present by the Ginsberg Company, and has been proved in several bore-holes on the property of the East Rand Proprietary Company.

The Kleinfontein, Van Ryn, Chimes, and Modderfontein Reefs, of which the mutual relations are somewhat obscured by a complex system of faults, consist of two reef-series (North and South Series) separated by about 600 feet of sandstone. The North Reef rests immediately on shales; it consists of from a few inches to 5 feet of conglomerate, carrying good-sized pebbles, a large portion of which are dark coloured. In the Kleinfontein property the greatest amount of gold is found in a leader on the hanging-wall side of the North Reef, some 12 to 20 feet to its south. The South Reef series consists of a largish body of low grade banket on the foot-wall, and a much smaller body (the rich Chimes or Modderfontein Reef) on the hanging-wall side. This reef is composed of two or more small leaders, each an inch or two in thickness, and separated by a variable amount of quartzite or grit; the hanging leader is often phenomenally rich. There is little doubt that this system of reefs, although separated by a considerable break from the central section, belongs to the Main Reef series, since, apart from their gold contents, the general trend of the formations (quartzite-shale formation on the north and Bird and Kimberley Series on the south) indicates a true connection with the Reef Series of the Central Rand. Farther east the beds can be followed through Modderfontein to Klipfontein, where they are now being tested by diamond-drill boring on the property of the Rand Klipfontein and Amatola Syndicate. Still farther east they are lost under the overlying coal-beds of the Karoo formation.

The Du Preez or Rietfontein Series.—The series of conglomerates known by this name occurs on the farm Rietfontein about 2½ miles north of the outcrop of the Main Reef Series on Driefontein; and although a vast amount of prospecting has been done in the hope of finding elsewhere similar rich ore-bodies north of the Main Reef Series, hitherto these efforts have not been successful.

On Rietfontein the series crops out for a distance of about 10,000 feet, nearly the whole length of the mynpacht, and can be traced eastwards to the adjoining farm Witkoppie. Speaking generally, the formation consists of alternating bands of quartzites and shales, with occasional beds of conglomerate and grit. Its northern margin abuts against an enormous body of igneous rock of diabasic character; and the beds in immediate contact with this intrusive mass are much disturbed and dislocated. Through the major portion of the property the strike of the beds is roughly parallel to that of the Main Reef Series, namely east and west; but at the western end of the mynpacht there is a sharp bend in the beds, turning them a direction almost at right angles to that previously maintained. This south-westerly strike brings the beds abruptly up against the quartzites that form the first ridge lying north of the Main Reef Series, showing plainly that the series must have been faulted at this point. The dip of the formation is to the south. A general section taken along a line running north and south discloses the following sequence. Beginning with the upper beds, there is first a series of well-defined but low-grade conglomerates known as the South Rietfontein Series. These beds crop out beyond the south boundary of the mynpacht, but are included in the

Rietfontein Estate Company.[1] Separated from them by some 2300 feet of shales and quartzites is the so-called "Stable Reef," which consists of a series of grits about 100 feet wide, with a small seam of conglomerate on the foot and hanging walls. Some 300 feet of quartzites intervene between the Stable Reef and the pay-leaders, which are two small seams of banket known as the Middle Reef and Middle Reef Leader. North of the Middle Reef, at a distance of 20 to 40 feet, is another small seam of banket known as the North Reef. North of this for a considerable distance the country is wholly diabase, and in places this igneous mass trespasses on the conglomerates and conceals their outcrop.

The North Reef averages about 8 inches in width, although in places it is considerably wider. It is enclosed within quartzite walls, that forming the foot-wall being interlaminated with a green talcose mineral. The North Reef has only so far been found payable in the western section of the property. The Middle Reef and its leader are narrow seams of conglomerate separated by a 2 to 3 feet seam of quartzite. The banket consists of pebbles of white quartz embedded in a hard quartzitic matrix. In consequence of the intense metamorphism which the beds have suffered, it is often difficult in the unoxidised ore to distinguish the pebbles from the matrix. The reef is, however, distinguishable by its high pyritic contents, and by the dark colour of the matrix. In places the leader is represented by an extremely narrow seam of pyrites in the form of small pellets of the size of buck-shot. The mine is celebrated for the very high returns of gold which from time to time it has produced, and for the rich specimens showing visible gold that have been frequently found. The gold in such specimens is generally dusted over a cleaved face, and is associated with small black spots which appear to consist

[1] Quite recently the ground of the S. Rietfontein Company has been amalgamated with that of the New Rietfontein Company.

of graphite. Work in the eastern section of the property has proved a valuable shoot of ore. Below the third level this high-grade ore was cut off by a strike fault, and some considerable exploration had to be undertaken before the faulted portion was recovered. Payable ore has also been worked to some extent in the western section of the property ; but the central portion is practically virgin ground.

Botha's Series.—The Botha's Series is generally taken to be the western extension of the Main Reef Series. But in order to give tangibility to this view, which, be it said, has powerful facts to support it, it is necessary to assume the existence of a big break between the eastern termination of this series and the western termination of the undoubted Main Reef Series. We shall have occasion to discuss this matter fully under the head of faults ; it is only necessary, therefore, to point out that the last properties in which the Main Reef Series is undoubtedly proved are those of the Banket and Gipsy Companies on the farm Witportje. The Teutonia, which is the first mine on the western side of the break, lies four miles north of these properties. East of the Teutonia the Botha's Series has not been traced, while westwards it has been followed through Witportje, Luipaardsvlei, and the south-east corner of Waterval, where it takes a turn to the south, running through Uitvalfontein, Randfontein, and Middelvlei. Beyond Middelvlei it appears to course in a south-westerly direction, but it has not yet been definitely located. The Companies working on this series are the following, taking them *seriatim:* Teutonia, Champ d'Or, Windsor (late Britannic), Luipaardsvlei Estate, York (late Emma), the George and May (late Botha's), and Randfontein Estate. The following section of reefs is generally characteristic of the district. The northernmost bed is a large body of banket usually described as the Main Reef. Some 20 to 40 feet to the south of this is one, or in some cases two small leaders, representing the South Reef. The Main Reef is of low grade, and as a rule is not worked, excepting where it is of payable character, as at the Champ d'Or. The leaders of the South Reef on the other hand are generally payable, and are being worked in the mines of the Luipaardsvlei Estate, George and May, and Randfontein.

Battery Reef Series. — This series may be traced at intervals through the farm Luipaardsvlei and Rietvlei, following the trend of the Botha's Series, and lying about 1500 yards to the south-east of the latter. Through the greater part of Luipaardsvlei the beds have an east and west strike, but in the western portion of the farm they turn to the south, and in Rietvlei strike almost due north and south. On Luipaardsvlei the series passes through the properties of the Luipaardsvlei Estate and Violet Consolidated Companies, while on Rietvlei it runs through the claims of the Horsham Monitor, Rietvlei, Vulcan (formerly the Vera), First Netherlands, and Lindum Companies.[1]

The Battery Reef Series comprises twelve or thirteen beds of conglomerate, some of which are of considerable size, but only one payable reef is known to exist in any one property, although it is not quite certain that the same member of the series constitutes the pay-body in all the properties. The pebbles of the pay-reef are of a fairly large size, and both hanging and foot walls consist of quartzite. The best portion of the reef is on the foot-wall, which is characterised by a layer of very large pebbles and by a honeycombed structure. Although the gold contents of the Battery Reef undoubtedly exist in the form of patches, there is no doubt that certain portions of the reef are of a payable character.

On the supposition that the Botha's Series represents the western extension of the Main Reef Series, the evidence in

[1] Other smaller companies of the Battery Reef are the Fern (now known as the Rip), the Old Edna (now part of the Champ d'Or Deep), Vanwyk, Lucas Bros., and Clarke and White's.

favour of which we shall discuss later, the Battery Reef Series must be correlated with the Kimberley Series of the Central Rand, while the intervening series opened on the Monarch and Standard properties probably belong to the Bird Reef Series.

The Black Reef.—The Black Reef[1] is the youngest auriferous deposit of the Rand formation. Its outcrop is met with at a distance of 6 miles south from the outcrop of the Main Reef Series. Although it strikes roughly parallel with the other conglomerate series, there is no doubt that an unconformity exists between the Black Reef formation and the underlying beds. The following facts may be adduced in support of this view. First, the Black Reef lies at a much flatter angle than the older beds, its dip rarely exceeding 10°. Further, if it is followed to the west, it is found to gradually overlap the older series, until on Middelvlei it probably even oversteps the outcrop of the Main Reef. In support of the same view is the fact first pointed out by Penning,[2] that the Black Reef beds are underlaid by a sheet of amygdaloidal igneous rock, which was probably formed out at some period intermediate between their deposition and that of the older beds. Another fact is the occasional presence in the Black Reef of rolled pebbles of older conglomerates. The work done on the Black Reef has been chiefly concentrated on the farm Roodekop, which lies due south of Elandsfontein or south-east of Johannesburg. The reef has been mined with considerable profit on the properties of the Orion and Meyer and Leeb companies. Other companies already at work, or in course of preparation for active work, are the Minerva (an offshoot of the Orion and Mulders Farm amalgamation), the Golden Kopje, Cornucopia, East Orion (late "Black Reef"), South Orion, and Pleiades Grosvenor. East of Roodekop the reef has been traced through the farm

[1] So called because of the presence of a black seam on its foot-wall.

[2] *Quart. Journ. Geol. Society*, xlvii. 1891, Plate xv.

Klipportje, Finaal's Pan, and Witpoort; while west of the same farm it has been followed through Palmietfontein, Rietvlei, Olifantsvlei, Misgund, South Luipaardsvlei, and Middelvlei. In the last-named farm the reef has been opened successfully in the properties of the Midas Black Reef and Howick and Rockley companies. We have already referred to the occurrence of the Black Reef in the Klerksdorp district (p. 19).

Like the older series, the Black Reef generally consists of a bed of conglomerate. The pebbles, which on foot-wall sometimes attain to the size of a hen's egg, consist mostly of white and smoky quartz, but to some extent also of a pinkish quartzite. The lower part of the reef is the most heavily mineralised, being charged with pyrites as well as containing nodules of that mineral. In width the conglomerate bed is very variable: in some places being as much as 6 feet; in others very small or even entirely absent. The most characteristic feature of the Black Reef is the highly ferruginous clay or slaty seam that forms its foot-wall and carries the greatest proportion of the gold-contents. The hanging-wall rock is invariably quartzite, while under the reef is a red clay, which possibly may have resulted from the decomposition of igneous rock. With regard to the gold-contents of the Black Reef, engineers are agreed as to its "patchy" or "spotty" character, that is to say it is not of uniform value throughout, the pay-ore being frequently limited to rich zones or patches of reef of varying dimensions. On Roodekop there appear, however, to be definite "shoots" of ore, having an east and west strike. Of these there are at least two, one traversing the Orion and the other the Meyer and Leeb property. They are about 300 feet apart and are roughly parallel. The Orion shoot has been worked for over 1000 feet along the strike, and has yielded excellent returns to that company. Other shoots no doubt exist, which will be discovered on active prospecting being undertaken. Although not such a certain mining proposition as the Main Reef Series, there are no doubt great possibilities in the Black Reef.

The Lanham and Vulcan Series.—Overlying the beds of the Witwatersrand formation proper in the west Rand is a flat or "blanket" formation consisting of hard white quartzites, underlaid by clays and shales. Between the quartzite and the clay is an irregular seam of conglomerate, at the bottom of which lies a dark ferruginous seam which in places carries gold. This formation is found on the farm Waterval a few miles west of Krugersdorp, where it forms a shallow basin having a very sinuous outcrop extending over several miles. The discovery of a rich patch in this bed caused a considerable amount of claim-pegging in this district last year, and some exploratory work and test crushings were made on the properties known as the Otto and the Queen. But the gold-contents proved too erratic for profitable working. There is no doubt as to the unconformity existing between this formation and the underlying beds of the Witwatersrand formation which dip at a higher angle, and outlying patches of the same series have been found in many places in the West Rand, for instance on Rietvlei and Randfontein. On Rietvlei the discovery of a rich patch resulted in the flotation of a company known as the Vulcan. The treacherous character of the reef, however, soon brought work to a standstill on this property. There can be little doubt that the Lanham and Vulcan Series are geologically identical with the Black Reef; as points of resemblance the clay underlay, the ferruginous seam, and the unconformity to the older beds of the Rand formation can be instanced.

Steyn Estate Series.—The reef worked by the Steyn Estate and Madeline Companies crops out in the northern margin of the farms Doornkop, Zuurbult, and South Luipaardsvlei. It is about 2 to 2½ feet wide, and consists of pebbles of white and smoky quartz. With the pay-bed are associated other beds of conglomerate and quartzite.

The ground is much disturbed, probably by the intrusion of igneous rock. In the Steyn Estate the reef has been tilted over at the surface so as to dip northwards, but it resumes its true southward dip in depth, after having been broken off by a dyke below the 140-foot level. In the present state of our knowledge it is impossible to correlate the Steyn Estate Series with any of the known members of the Witwatersrand conglomerates.

Dykes and Faults

Like any other stratified formation (say a coal-field) that has been tilted from its horizontal position, the Rand

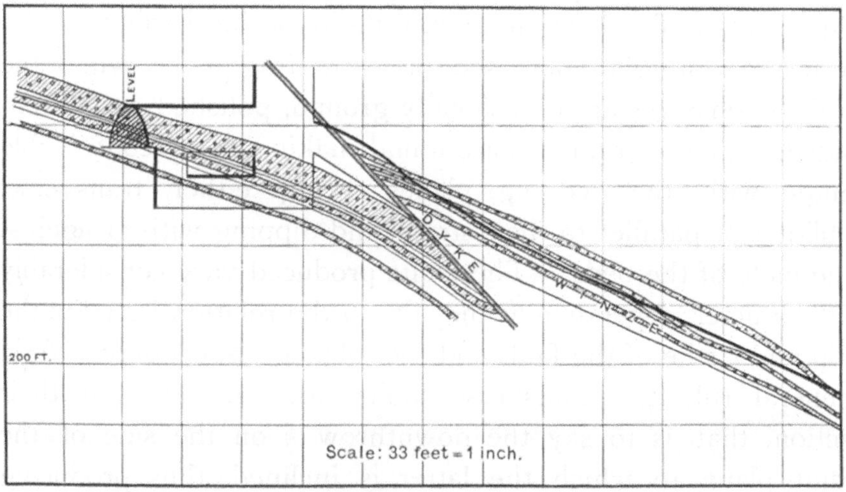

Geological Cross Section in Simmer and Jack Mine.

Fig. 43.—Example of Fault and Dyke.

beds have been fractured, dislocated, and pierced by intrusive masses of igneous rock. The disturbances in stratification caused thereby, which among mining men are roughly classified as faults and dykes, play a considerable rôle in Rand mining, partly on account of the difficulties they give rise to in developing and stoping, and partly by reason of the actual loss of tonnage incurred by the substitution of barren rock for payable ore-bodies. So much so is this the case that a mining engineer, in estimating the ore reserves of a Rand mine, generally considers it necessary to make a reduction of 10 to 15 per cent on the calculated quantities in order to allow for losses which may be incurred on this account. In view of the misapprehension that prevails on the Rand and elsewhere as to the true nature of these disturbances, it may be well to take this opportunity of explaining what is meant by the terms fault and dyke when used scientifically. A fault is a dislocation or displacement of the beds due to differential movements of portions of the formation while undergoing upheaval. The plane along which the fractured walls have slipped is termed the fault plane. It is usually marked by a seam of clay (flucan) produced by the fine comminution of the opposing surfaces, or by a friction breccia composed of large fragments, while the walls themselves are occasionally ground, polished, or slickensided. The fault planes are found making every conceivable angle with the bedding planes, being either transverse, oblique, or parallel to their strike, and dipping with or against the hade of the beds. The faults produced vary considerably in character, depending as they do on the relation between the dip and strike of the fault and reef planes respectively. As a general rule the transverse faults are "normal" in their action, that is to say the downthrow is on the side of the fault plane to which the latter is inclined, thus producing "step faults," while the strike faults, or those that cross the beds at an acute angle, are frequently "reversed" or "overthrust" faults causing an overlap or reduplication of the beds.

Dykes, on the other hand, are wall-like bodies or sheets of igneous rock that have been intruded into the formation upon some plane of weakness which may be either a simple fracture or a fault plane. In the former case the dyke does not throw or heave the beds, in the latter it does. In some instances the igneous matter has been intruded along bedding planes, in which case the dyke will be found to dip and strike with the formation. In the great majority of cases, however, the dykes cross the formation either at right angles or obliquely, and their walls usually approximate to the vertical. On the lower levels of a mine the igneous rock is found in its unaltered hard condition, but on the upper levels it is decomposed, being replaced by a soft red or white clay resulting from the kaolinisation of the felspathic constituents of the rock.

Petrologically these igneous rocks belong to the group of dark-coloured basic greenstones among which the following types have been recognised:—diabase, olivine-diabase, bronzite-diabase, epidiorite, gabbro, and olivine-norite. These rocks are composed of the usual minerals, namely:—plagioclase-felspar, augite, hornblende, bronzite, olivine, quartz, apatite, sphene, epidote, chlorite, and calcite, and have the micrographic structure characteristic of each individual type. In many instances the dykes found traversing the Rand formation are of a slaty or schistose character, which in some cases has prevented their being recognised as igneous rocks. This structural modification of the original igneous rock has been produced by a mechanical metamorphism due to movements of the earth's masses during faulting and upheaval. In such rocks the original augitic and hornblendic minerals are replaced by chloritic, talcose, steatitic, and serpentinous metamorphic products.

In the following pages we give a few examples of the more important dislocations and igneous intrusions occurring on the Witwatersrand goldfields.

Both east and west of Johannesburg, at about the same distance, a considerable break occurs in the normal line of the formation. The western break occurs on Witportje about 17 miles west of Johannesburg, while the eastern break is near Boksburg 15 miles east of Johannesburg.

The Witportje Break.—As already pointed out on page 35, the Main Reef Series is continuous to the Princess, but shows signs of considerable disturbance on the properties of the Banket, Bohemian, and Gipsy Companies. The exact location of the reefs still farther west is at present the subject of investigation. A number of reefs and detached portions of reefs crop out on the southern portion of Witportje, and on the northern portion of Vlakfontein, a farm lying south of Witportje. Several bore-holes are now being sunk on the northern margin of Vlakfontein, and these will doubtless furnish interesting information. Whatever may be the results of these investigations, there is, we think, very little doubt that at some point west of the Gipsy the conglomerate beds are cut off by a great fault which shifts the outcrop some 4 miles or more to the north. The facts in support of this view are the following:—

1. The westerly or south-western strike of the Main Reef Series would, if there were no fault, carry the outcrops through Rietvlei.
2. In the western portion of Rietvlei, however, we find a series of conglomerates (Battery Reef Series) striking north and south, and dipping towards the east. About a mile west of the Battery Series is the Randfontein Series, having a similar strike and dip.
3. The Randfontein Series followed north can be connected without much difficulty with the Botha's Series, and the latter assuming an easterly strike on the farm. Waterval can be traced through Luipaardsvlei and Witportje, where it is being worked by the Champ d'Or Company. East of the Teutonia, which adjoins the Champ d'Or on the west, the series is apparently cut off.
4. The members of the Botha's Series resemble the Main Reef Series, being composed of a big low grade body (Main Reef), and a rich south leader, some 20 feet or so to the south.
5. The Battery Reef Series must then be correlated with

the Kimberley Series, while the Monarch and Standard Reefs probably represent the Bird Reef Series.

The Boksburg Break.—To about the centre of the farm Vogelfontein and Main Reef Series is clearly recognisable, a payable body of banket having been opened on the properties of the Comet, Cinderella, and Blue Sky Companies. Beyond the Blue Sky, the series passes under the coal deposits of the Karoo Formation, and has not been farther traced. The Bird Reef Series crops out near the Boksburg Railway Station, some 2000 feet south of the Blue Sky Reef, while the Kimberley Series, lying another 3000 feet to the south, are found in Leeuwpoort, and have been opened on the properties of the Ziervogel and Leeuwpoort Companies. The eastern extension of these series is interrupted by the intrusion of a considerable mass of igneous rock (olivine-diabase), forming the elevation known as One Tree Hill. The northern formation (quartzite and shales), however, can be followed uninterruptedly through the farms Klipfontein and Kleinfontein; and consequently there can be no doubt that the ore-bodies worked by the Chimes Group of Mines constitute a true extension of the Main Reef Series. The farthest point west at which these reefs can be recognised are the outcrops occurring on the properties of the Spartan (now West Kleinfontein) and Florence Syndicates on the farm Benoni; but, beyond a few shafts sunk in the early days in these properties, there has been practically no work done to prove their value. We believe that in this respect a more energetic policy is now being pursued. The extent of the gap existing between the proved portions of the Main Reef Series is about 3½ miles. This space is occupied by the farm Rietfontein, the property of the Apex Syndicate.

The reefs worked by the Kleinfontein, Van Ryn, Chimes, and Modderfontein Companies, are much complicated by a series of transverse faults, cutting the ground up into blocks which have undergone considerable displacement relatively to one another. As stated above, however, a careful study of the district has elicited the facts that these apparently numerous reefs may be reduced to two series, some 600 feet apart, each carrying one valuable ore-body; while the Bird and Kimberley Series are also recognisable at their proper horizons.

Next in magnitude to the main breaks just described are the faults which occur on the properties of the

Simmer and Jack and *Geldenhuis Estate.*—Here two faults have been caused by the upheaval of a block of ground a mile and a quarter long. The superficial effect of this pair of faults are the breaks in the outcrop on the western boundary of the Geldenhuis Estate, and in the eastern portion of the Simmer Mynpacht. The first of these has a throw of 870 feet to the south, from which point the reef outcrop continues to the Simmer fault, where it is thrown back to its original line, the horizontal displacement here being 650 feet, measuring at right angles to the strike of the reef. The Geldenhuis fault crosses the formation as nearly as possible at right angles, while the Simmer fault crosses in a south-easterly direction, making an angle of 37° with the strike of the reef. As the Simmer fault dips to the north-west, there is here in all probability an overlap of the ore-bodies. The terminal points of the outcrop, measuring along the trace of the fault at the surface, must be at least 1000 feet apart.

Another fault of considerable size is that existing between the *Metropolitan* and *Henry Nourse Properties.* In the Metropolitan ground this fault is first met with near the boundary with the Doornfontein Mint, where the outcrop is cut off and thrown in a south-easterly direction into the ground of the Henry Nourse, the trace of the fault plane at the surface making an acute angle with the strike of the reefs. The actual displacement at right angles to the strike of the reefs is about 550 feet, but, measuring along the fault, the apparent displacement amounts to as much as 1500 feet. The dip of the fault is to the north, so that although the roofs do not outcrop on the ground of the Doornfontein Mint, or in the eastern section of the Metropolitan, which lies to the south of that property, they will be found in depth. This fault has been struck in one of the Henry Nourse deep-level shafts, the intersection setting free a considerable amount of water.

Good examples of over-thrust or reversed faults occur in the Robinson, Meyer and Charlton, Crown, Langlaagte Royal, Knight's Tribute, Glencairn, and Chimes mines. At *The Langlaagte Royal* the reef is thrown up three successive times between the outcrop and the seventh level by parallel east and west faults. The gain in backs caused by these overlaps amounts to some 400 feet. At the Meyer and Charlton the backs gained between two levels by reverse faulting amount in some places to 50 feet. The most striking example of the reduplication caused by this type of faulting is to be seen on the property of the Witwatersrand Company (also known as Knight's), where the conglomerates of the Main Reef Series have a double outcrop throughout the whole length of the property. The two lines of outcrops which in the central section are about 400 feet apart, show identically the same sequence of beds, each easily recognisable by its distinctive petrological character.

The position of the fault is marked by a dyke some 70 feet in width. Bore-holes put down to the south of the southernmost outcrop failed to find the supposed north reef, but, on being carried deeper, intersected this dyke. The gain in backs due to this fault is of course very considerable.

The strike fault existing in the eastern section of the property of the new Rietfontein Company is also of considerable interest on account of the important influence its discovery exercised on the market value of the shares.[1] The

[1] Before the influence of this fault was felt in the mine the shares stood at

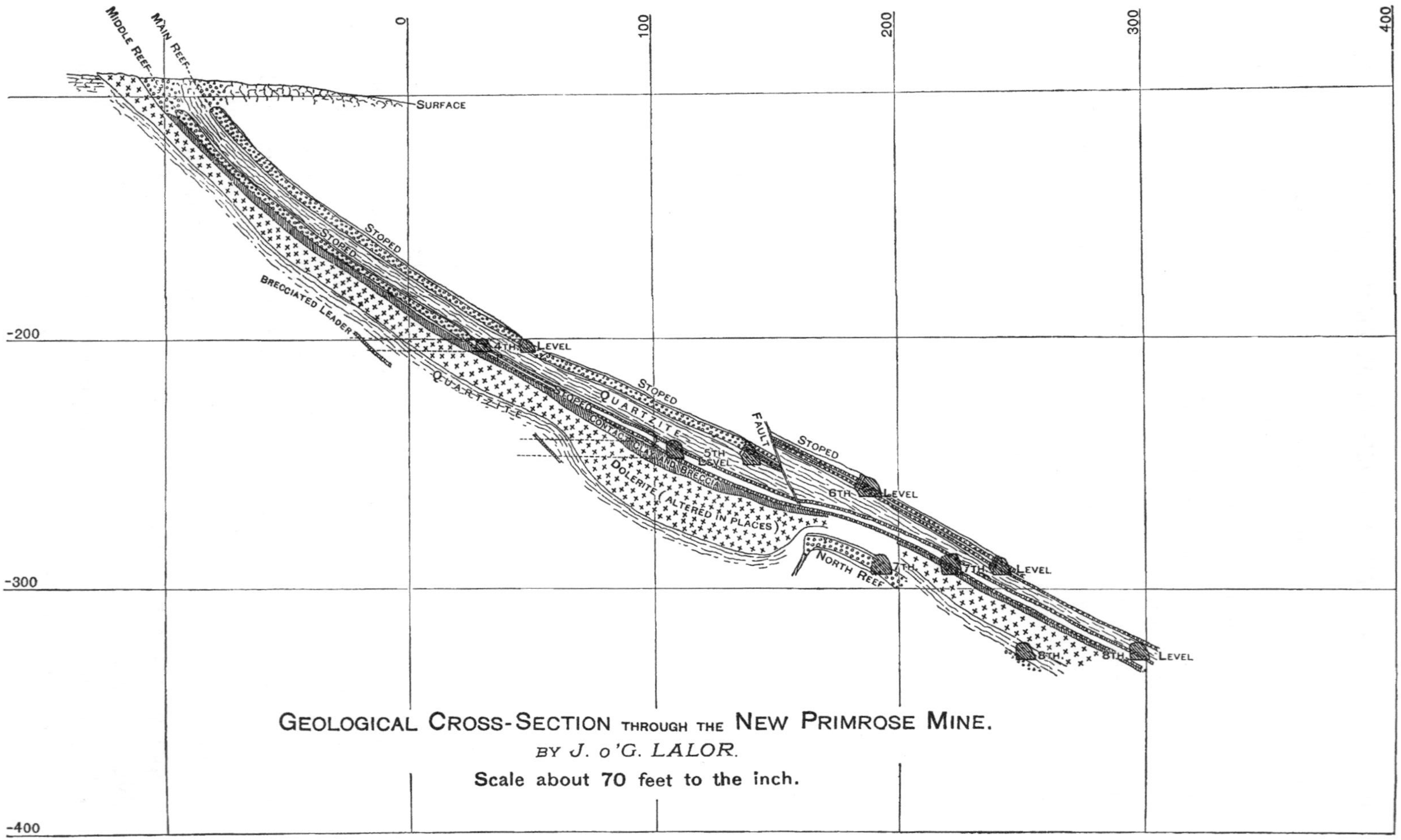

Geological Cross-Section through the New Primrose Mine.

by J. o'G. Lalor.

Scale about 70 feet to the inch.

Fig. 44.

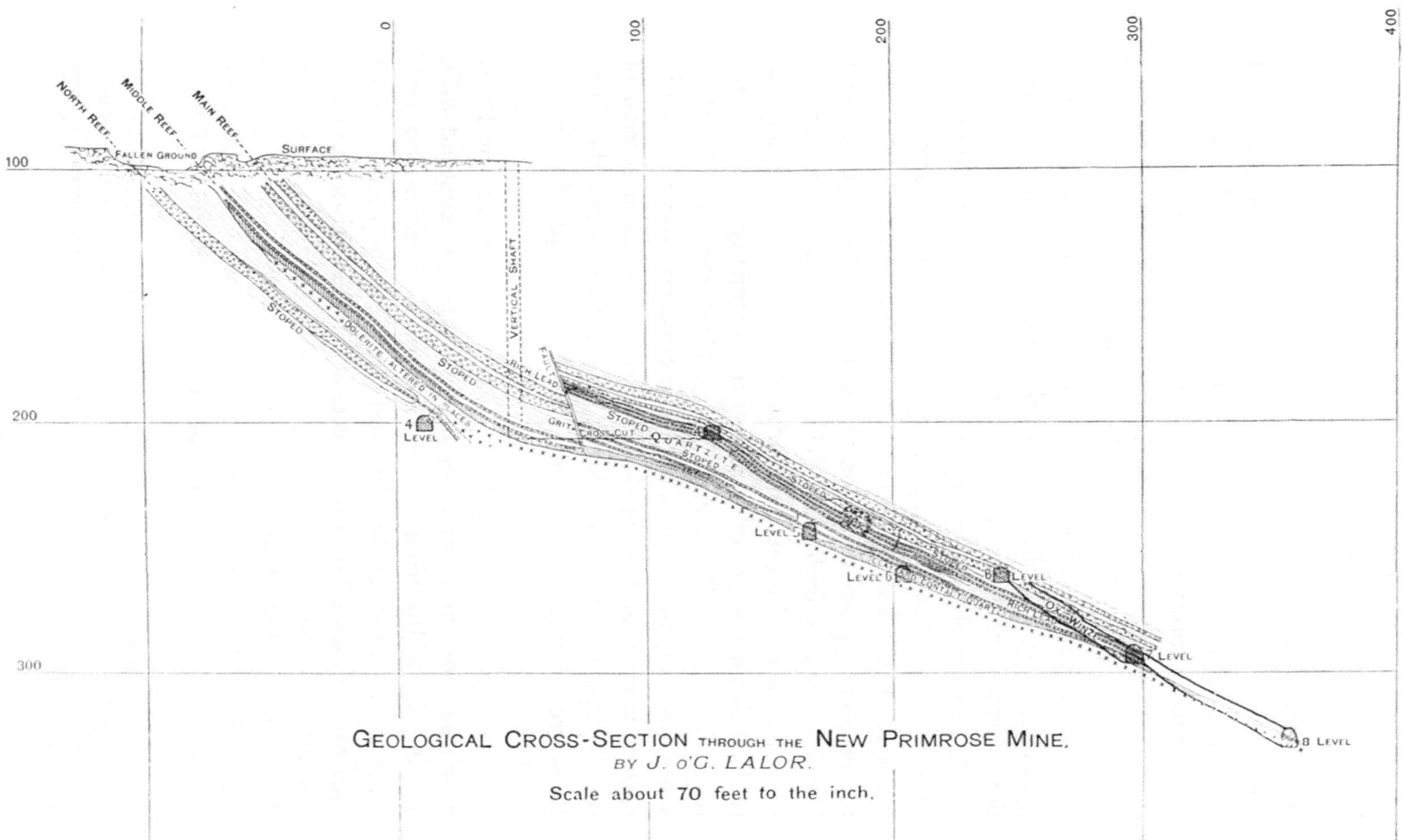

GEOLOGICAL CROSS-SECTION THROUGH THE NEW PRIMROSE MINE.
BY J. O'G. LALOR.
Scale about 70 feet to the inch.

FIG. 45.

fault which courses almost parallel to the reef dips at a slightly steeper angle. Its position in the mine is marked by a thin seam of white clay, and by the fissile character of the quartzite forming its walls. On following the fault down, the reef was picked up some 300 feet lower, the fault belonging to the normal class. The downthrow of this fault amounts to 300 feet, but the reef is again thrown up by two parallel upthrow faults, so that the loss occasioned by the downthrow is considerably reduced.

A considerable amount of disturbance by faulting has been discovered in working the group of mines known as the Main Reef, Unified, and the Aurora. The centre of the disturbed area appears to be located in the Unified, where there is quite a nest of faults and dykes; the difficulties in working the mine caused by these disturbances have had a great deal to do with the chequered career of that company. In the property of the Spes Bona and George Goch there is a break of some moment caused by a big dyke crossing from the south-eastern portion of the Spes Bona into the eastern section of the George Goch. The throw of the fault from the Spes Bona ground into the George Goch is about 300 feet northward.

With regard to dykes, there is hardly a mine on the Rand which has not one or more of these troubles to contend with, varying in width from a few feet to more than 100 feet. Once their extent, strike, dip, and effect on the reef is known they can be tackled on each successive level with a minimum amount of dead work, and any loss of tonnage caused by them can be allowed for in making estimates. Examples of big dykes may be studied in the Ferreira, Robinson, Wolhuter, Spes Bona, George Goch, Geldenhuis Estate,

£7 : 10s. As the original shoot which existed in this part of the property was gradually worked out down to the fault, the shares fell to 15s. On the reef being struck below the fault they at once rebounded, and have again passed £5 during the present year.

Simmer, Knight's Tribute, Unified, Champ d'Or, and Rose Deep.[1] In isolated instances the igneous matter has been intruded between the beds instead of breaking across them. An example of this kind occurs at the Wolhuter, where below the first level a thin intrusive sheet has forced its way between the Main Reef and the Main Reef Leader. As depth is made, the width of this sheet increases until on the third level the reefs are separated by 140 feet of igneous rock. Another interesting instance of an intrusive sheet occurs in the Primrose mine. Between the north and middle reefs is a sheet of igneous rock, swelling here and there into thick lenticular masses which squeeze out the north reef but do not disturb the middle reef. The greatest proved width of this sheet is 160 feet on the sixth level, but the width varies considerably, and in places the dyke thins out altogether. In those places where the dyke is absent the north reef is of good width and quality. The latest developments appear to show that this sheet is thinning out in depth.

[1] *Rose Deep First Annual Report*, 1894.—"No. 2 Shaft was sunk in dyke matter to a depth of 335 feet from the surface. For the first 100 feet this dyke was very soft and full of water, necessitating the driving of spiling ahead of the excavation for a considerable distance."

CHAPTER IV

PARTICULARS OF THE RAND DEPOSITS

1. Nature of the Mineralisation

As already stated, the "banket" beds are composed mainly of pebbles of various size consisting of pure white or grey quartz. These pebbles lie imbedded in a matrix consisting originally of sand, but now completely cemented to an almost homogeneous material by a later deposition of silica chiefly in the form of crystalline quartz. This introduction of mineral matter in solution, combined with great pressure, has caused the metamorphism of beds, originally existing as loose friable sandstones and shingle deposits, into hard compact quartzites and conglomerates. In places the excess of silica has segregated out, filling fissures and cracks, and forming small quartz veins. In other places the metamorphism takes the form of a schistose structure, or is indicated by presence of talcose seams. In such places too the pebbles are often cased by thin silvery films of mica and talc. Where the crushing has been great, the pebbles are occasionally broken and recemented, or their margin is partially obliterated, a portion of the pebble passing imperceptibly into the quartzose matrix. The pebbles do not as a rule carry any metallic constituent, the mineralisation being confined to the matrix. In those cases, where gold has been observed in the pebbles, it has probably been subsequently introduced along some small fissure.

Besides gold, the metallic minerals observed in the banket deposits are iron pyrites, marcasite, copper pyrites, blende, galena, hematite, ilmenite, and magnetite. Arsenical pyrites does not appear to be present in these ores; while antimony (in the form of stibnite) is of rare occurrence. Cobalt and nickel have been proved chemically. Among non-metallic oxides, rutile, zircon, and corundum have been detected by microscopic examination of the heavy residue obtained by gravity separation by the aid of the "heavy solutions" well known to mineralogists. The presence of corundum has been proved chemically by Mr. Crosse of the Standard Bank in Johannesburg; and the remarkably rapid wear in the battery mortar boxes is probably attributable in some degree to the presence of this mineral. The minerals muscovite, chlorite, and talc have also been recorded. They have probably resulted from the decomposition of some original felspathic constituent of the conglomerates.

By far the greater proportion of the metallic mineral constituents found in the unoxidised ore or "blue rock" consists of ordinary iron pyrites, amounting probably on the average to 3 per cent of the ore. This mineral occurs in small crystalline particles usually finely disseminated through the ore. Sometimes, however, it is aggregated to patches or nodules, or occurs in thin seams or streaks. In the oxidised ore occurring in the upper levels of the mines the pyrites is replaced by iron oxides (magnetite, limonite, and hematite) forming the so-called "black sands," resulting from the decomposition of the pyrites under the influence of the atmosphere.

The gold occurs in the metallic condition, and is only to a small extent alloyed with foreign metals. The fineness of the gold varies somewhat in different parts of the Rand, but is generally fairly uniform in one and the same mine.[1] As a

[1] A curious, and as far as we know unique, instance of an improvement in the quality of gold in depth has been recorded from the Pioneer Mine. A number of

rule it is disseminated through the ore in finely divided particles, which are only rarely of sufficient size to be visible to the eye, unless the ore has been previously crushed and the bulk of the quartz sand removed by washing in a prospector's pan.

Besides "free gold," by which is meant that recoverable by amalgamation, there is also present in the unoxidised ore a certain proportion of the noble metal which in practice is only extracted by chemical treatment. This "fixed" gold, however, is not in chemical union with any other element, for it has been clearly demonstrated by microscopic research that it is only mechanically included in the pyrites, and thus withdrawn from the action of the mercury. In the pyrites crystals there are a number of minute cavities, fissures, and structural planes in which the gold exists in minute particles and films. Mr. Crosse has in his possession some slides prepared for the microscope, showing the sub-crystalline character of the gold liberated from the enclosing pyrites by the action of nitric acid.

bullion assays, made by Mr. Crosse, of the Standard Bank, Johannesburg, extending over a period of nine months, showed the following variations in fineness :—

Parts per 1000.

Gold.	Silver.	Other Impurities.
816·5	145	38·5
835	145	20·0
841	142	19·0
833·5	140	16·5
832	140	28·0
829	141	30·0
833	141	23·0
834·5	140	25·5
826·5	136	37·5
829	136	35·0
834·5	137	28·5
840	131	29·0
827	121	52·0
840·5	126	33·5
844	124	32·0
846·5	121	32·5
854	121	25·0
858·5	118	23·5

2. Origin of the Ore-Bodies

It is now believed by many that in the banket beds, as in quartz veins, the gold is of subsequent origin, not having been deposited in the form of small particles and dust by the waters that were instrumental in heaping up the shingle or conglomerate beds. The theory that the gold is of primary origin had at one time many supporters (Zirkel, Pelikan), but has now gone out of favour. The different ways in which the mineralisation could have been effected continues, however, to occupy the attention of theorists. According to some the gold has been deposited from ascending or descending thermal solutions (Koch, Curtis, Rathbone), others believe that it has been sublimed in the form of vapour; and lastly, a few suggest that the auriferous wealth of these deposits has been precipitated chemically from the seas that formed them (Stelzner).

We do not propose here to discuss the *pros* and *cons* of these various hypotheses; we will, however, contribute a few geological facts bearing on the origin of the banket deposits. With regard to their lithological characters, the banket deposits are nothing more or less than ordinary conglomerates; and conglomerates, geology teaches us, are marginal sea deposits that have been formed from the waste of pre-existing rocks. These older rocks were probably members of the Primary formation, granites and schists, on which the Witwatersrand beds lie. That they were largely veined with quartz is evident from the nature of the pebbles, and that they were *not* the source of the gold contained in the banket deposits is also evident from the fact that the quartz pebbles do not as a rule carry gold.

The shingle-beds or conglomerates were no doubt deposited in practically horizontal strata. The repetition and alternation of conglomerate beds with those of sandstone is explained by the alternating slow upheavals and subsidences of the sea-bottom, which geologists have proved must have taken place, and in fact are taking place at the present moment. The inclined position in which the beds lie at present is due to the tilting of the formation which was occasioned, partly by movements of the earth's crust, and to some extent, perhaps, by the intrusion and upheaval of igneous masses.

The displacement of such enormous masses of rock naturally caused the formation of planes of disruption and sliding transverse to and parallel to the bedding. This fissuring of the formation often led to a considerable faulting of the formation, blocks of ground being raised or depressed in relation to the neighbouring blocks, and the upper beds pushed over those underlying them, producing overlaps and reduplications. Not unoften the conglomerate beds themselves were coincident with the planes of movement, as is evidenced by the seams of clay ("flucan") and vein-quartz, which frequently accompany them. The formation of faults and fissures was followed by injection of molten igneous matter, giving rise to the dykes which are such a familiar feature of the mines. As is well known to those who have had occasion to study these phenomena, some of these dykes are found to fault the beds, while others cause practically no disturbance. The reason of this difference is clear: in the one class of dykes the igneous rock was intruded along a fault plane, in the other class along a simple plane of fissuring. No doubt the same planes acted as channels for the introduction of ascending mineralising solutions, either contemporaneously with or subsequently to the igneous intrusions. In some cases the dykes were evidently formed prior to the period of mineralisation, since the conglomerates often show a considerable difference in gold contents on the opposite sides of a dyke. We have not been able to find any reliable evidence in favour of the idea locally prevalent that the dykes have acted beneficially on reefs in their immediate neighbourhood in regard to gold-contents.

The fact that the conglomerate beds are mineralised in preference to the sandstones and quartzites is due no doubt to their greater permeability on the one hand, and in part also to the fact, already noted, of fissuring and sliding having taken place in the plane of those beds.

3. Distribution and Value of the Gold-Contents of the Reefs

That the distribution of the gold in the Rand banket is much more regular and uniform than in most other auriferous deposits is evident from the continuity of the areas from which the ore has been taken out in the majority of the mines that have been working for several years. A glance at the stope plan or projection of almost any profit-saving Main Reef mine will show that the proportion of absolutely worthless patches of reef is very small when compared to the ratio obtaining between pay-ore and barren ground in the general run of quartz or vein deposits. The transition from a rich to a poorer quality of ore in the banket mines is, generally speaking, gradual, and rich sections are not separable from those of lower grade by any well-defined line of demarcation, such as is characteristic of most other metalliferous deposits. The banket deposits are generally described as containing their gold rather in "patches" than in "shoots." This is true when the banket reefs are studied in detail. When, however, the deposit is considered as a whole, the prevalence of a "patchy" character on a large scale is not quite so evident; there are instances, moreover, of distinct shoots, such as that existing in the eastern portion of the New Rietfontein property, that occupying the central portion of the Simmer and Jack Mine, and the parallel strike-shoots of the Meyer and Leeb and Orion properties.

On studying the Main Reef Series as a whole, we find

that there are certain sections of higher grade than others, which as far as present developments have been carried, cannot be described either as shoots or patches, since their lateral extent is not disproportionate to the depth to which they have been proved to maintain their richness. As examples of such sections we may instance the central portion of the Rand, embracing such high-grade mines as the Robinson, Worcester, and Ferreira; the Heriot and Henry Nourse section; the Chimes and Modderfontein section; the Durban Roodepoort; the Champ d'Or and the Nigel.

The very high assays, amounting to several hundred ounces to the ton, which occasionally appear in the companies' reports, and can be obtained in sampling almost any mine on the Rand, are due to the presence in the banket of small sporadic patches or "spots" of exceedingly rich ore. Another cause of such abnormally rich samples is the presence of narrow seams, streaks, or leaders of rich ore, which sometimes occur on the hanging or foot-wall of the reefs. In the early days of the fields the rich crushings obtained in the small five and ten-stamp mills then in use were got by mining these rich leaders and by picking the ore from the rich spots and patches. As the mills increased in size so the grade of the ore fell off; and even at the present time the addition of more stamps to a mill is generally followed by a decrease in the grade of the ore milled, since ore-bodies which hitherto were regarded as unpayable become profitable under improved economic conditions. At the same time profits are increased by the reduction of working costs consequent upon operating upon a larger scale.

Turning now to the respective value of the different reefs that make up the Main Reef Series, we find that here also striking differences are observable, since there are reefs of a highly profitable character, others of medium grade, and again some that are unpayable. Further, it is to be found that though one and the same reef may be sometimes remarkably uniform in value over considerable stretches of country, there are also variations in this respect. Thus in the central section near Johannesburg, we find the Main Reef Leader and the South Reef having a high value, while the Main Reef is of low grade though probably payable.

In the mines lying to the west of the Langlaagte Estate, chief dependence is placed on the South Leader, the Main Reef Leader being of irregular character and the Main Reef poor; while in the eastern section of the Rand the grade of the South Reef falls off, and the ore-bodies associated with the Main Reef become the chief producers.

It is very difficult to arrive at the value of the different ore-bodies worked in different mines, except in those few instances where test-crushings have been made of each reef separately. Of course an approximate idea of the value of a reef stoped to a given thickness can be got by careful sampling in the mine; but sampling is such a delicate operation, requiring so much skill and judgment, that the results obtained are only reliable in the few cases where special attention is paid to this department.

The following particulars relating to ore values have been taken in the main from the companies' published reports:—

The Robinson.[1]—In this mine the Main Reef, sampled over 2070 feet of drives, gave an average value of 10 dwts. 15 grains for an average width of 3 feet 2 inches. A test-crushing made on 1250 tons yielded—

	5	dwts.	15.3	grains	in the mill,
leaving	1	,,	0.42	,,	in the concentrates,
and	1	,,	18.50	,,	in the tailings.
Total . .	8	,,	14	,,	

The Main Reef is now being mined and mixed with the richer rock derived from the South Reef and Main Reef

[1] See *Annual Report* for 1894.

Leader stopes. The proportion in which these reefs are milled is stated in the Company's Reports to be as follows:—

	1893.	1894.
South Reef . .	30 per cent.	37 per cent.
Main Reef Leader	59 ,,	48 ,,
Main Reef . . .	10 ,,	15 ,,

The average mill yield was about $1\frac{1}{2}$ dwts. less in 1894 than in 1893, having been 22.11 [1] in 1893 and 20.56 in 1894 The average ore value of the total rock crushed in 1894, viz. 107,935 tons, was between 26 and 27 dwts. In 1893 it was 27·64.[2]

The Ferreira.—This is one of the highest-grade mines on the Rand, the South Reef being very rich and the Main Reef Leader of unusually high value. In the eastern portion of the mine the Main Reef and the Main Reef Leader are some distance apart, and the Main Reef is not mined; but in the western section they are in close juxtaposition, and a portion of the Main Reef has to be taken with the Leader, about 3 feet of the two reefs being stoped together.

The following values are taken from the General Manager's Report for 1894:—

Main Reef Leader.			Main Reef Leader and Portion of Main Reef mined with it.	
Level.	Average Width.	Value per Ton.	Average Width.	Value per Ton.
Feet.	Inches.	Oz. dwts.	Inches.	Oz. dwts.
320	6.43	5 : 4.27	...	...
420	12.0	1 : 5.90	36	0 : 16.65
520	15.72	2 : 9.67	39	1 : 5.63
620	26.25	2 : 2.25	44	1 : 7.86
720	5.21	2 : 5.43	...	...

[1] The yield for mill, concentrates or tailings, is given in bullion in the Chamber of Mines Returns. The fineness of plate gold averages about 880, that of chlorination gold 960, and that of cyanide gold about 700.

[2] *Vide* the "Report of the Concentration Committee of the Chamber of Mines."

South Reef.			
Level.	Average Width.	Value per Ton.	
Feet	Inches	Oz.	dwts.
420	16·58	3	15·55
520	18·10	4	14·26
620	25·08	5	10·40
720	24·00	3	2.75
820	36·00	3	14·58

The average mill grade for the year 1894 was 21.92 (bullion). Taking the last nine months of the year, 66,543 tons of ore were mined, having an average value of 19.33 dwts. (fine) per ton. By sorting out 44.72 of waste rock, the value of the ore was raised to 33.76 dwts. (fine) per ton (General Manager's Report for 1894).

The Wemmer Mine.—The ore mined during the last six months of 1894 was 42,673 tons, of which 81 per cent was from the South Reef, the remaining 9 per cent being from the Main Reef Leader. The Main Reef was not touched. Of the ore mined 35·14 per cent was picked out and rejected as valueless. The manager, Mr. Johns, anticipates that towards the end of 1895 he will, by improving the sorting plant, be able to discard, as waste, 45 to 50 per cent of the ore mined, thus improving the grade of the mill-rock in a corresponding degree. The following particulars are excerpted from a table given in Mr. Johns's Report for 1894:

Period.	Total fine gold in ore treated.			Amount of fine gold won by amalgamation.		
	Oz.	dwts.	grs.	Oz.	dwts.	grs.
Half-year ending 28th Feb. 1893	1	1	12.513	0	15	10.289
Half-year ending 31st Aug. 1893	1	2	15.497	0	15	14.136
Half-year ending 28th Feb. 1894	0	17	16.092	0	11	17.245
Half-year ending 31st Aug. 1894	0	17	3.95	0	10	17.33

The Main Reef Leader will probably be found payable throughout in the lower levels, and a greater percentage of this reef will be mined when the stamping power is increased.

In the *Salisbury*, *Jubilee*, *Treasury*, and *Village* Group of Mines, the South Reef is the staple producer, the Main Reef Leader being on the whole of low grade; while the Main Reef, which is in close juxtaposition to the Main Reef Leader, is considered to be unpayable except perhaps in the Jubilee. The South Reef for a 27-inch stope probably averages about 1 ounce of fine gold per ton, of which some 12½ dwts. are recoverable on the plates. The Main Reef Leader for the same stoping width has an ore value of about 12 dwts. fine, yielding some 8 dwts. on the plates. The lower 3 feet 6 inches of the Main Reef is said to have an assay value of 7 to 8 dwts. in the Jubilee. The Main Reef and the Main Reef Leader, which were for a time worked together in the Treasury after the extraction of the South Reef, yielded 5 dwts. on the plates. In the Salisbury the South Reef and the Main Reef Leader have in the past been worked in the ratio of 3 tons of the former to 2 tons of the latter. The average yield from all sources since this mine started, up to the end of 1894, was 23.97 dwts. per ton, consequently the original value of the ore must have averaged about 28 dwts. per ton. In 1894 the average plate yield was 11.62 dwts., it follows that the gold-contents of the ore mined during that period was about 17 dwts. per ton. The cause of this decrease in value is said to have been due to an actual falling off of the ore in certain levels.

The City and Suburban.—During 17 months (Jan. 1893 to May 1894) 12½ per cent of the total rock milled was from the Main Reef Leader; the remainder was South Reef, no Main Reef being mined in this property. The average yield per ton by amalgamation and concentration was 14.74 dwts. fine per ton. The average assay value of the tailings running from the mill was 5.73 dwts., making a total ore value of 20.47 dwts. per ton. Of this the South Reef probably averaged about 27 dwts. per ton, while the Main Reef Leader averaged about 12 dwts. During this period a 50-stamp mill was running, crushing under 5000 tons per month. The results now being obtained with a 130-stamp mill, crushing 15,000 tons per month, are of course much lower, the plate yield being about 9 dwts., which is equivalent to an ore-value of about 15 dwts. This difference, which invariably follows the replacement of a small mill by a large one, is chiefly due to a larger proportion of Main Reef Leader being mined than formerly, and to less discrimination being observed in stoping the reefs, low-grade rock, which formerly would have been left in the stopes, being now within the margin of profitable mining.

Meyer and Charlton.[1]—The reefs worked are the Main Reef, Main Reef Leader, and South Reef,—the South Reef and the Main Reef Leader both being of good value, while the Main Reef, according to recent tests, appears also to be a payable ore-body. The reef is 5 feet to 12 feet in thickness. Good results were obtained from a milling test made on 56 tons of ore of Main Reef, mined over a width of 12 feet on the fifth level. The average assay value from truck samples was 13 dwts. per ton. The gold won from the plates was 6 dwts. 20 grs. per ton, the tailings assaying 5 dwts. 8 grs. per ton.

The Crown Reef.—The reefs being worked in this property are the South Reef and the Main Reef Leader. The Main Reef, which is a big body some 8 to 10 feet in width, is not considered payable, its average assay value being stated to be about 4 dwts. According to the General Manager's Report for the year ending 31st March 1895,[2] the then developments on the Main Reef Leader and South Reef showed the following ore values :—

1 *Sixth Annual Report* to 31st Dec. 1894.

2 *Seventh Annual Report* for the year ending 31st March 1895.

On the fourth level (346 feet vertical), the average width and grade of the reefs as disclosed in the drives is as follows:—

	Width.		Grade.	
	Feet	inches	Ozs.	dwts.
Main Reef Leader . . .	1	3	1	13
South Reef . . .	2	6	3	5

On the fifth level (vertical depth 445 feet), 1736 feet were driven during the past year on the Main Reef Leader, and 1757 feet on the South Reef. The reefs exposed show the following average assay value: the Main Reef Leader for an average width of 1 foot 10 inches assays 2 ozs. 6 dwts.; and the South Reef, for an average width of 2 feet 2 inches, assays 2 ozs. 6 dwts.

On the sixth level (vertical depth 540 feet), the Main Reef body, where cut in the main cross-cut, showed a width of 17 feet, including 3 feet of Leader next to the hanging-wall. The Leader assayed 1 oz. 3 dwts. to the ton, but the main body (Main Reef) averaged only 2 to 3 dwts. per ton.

The seventh level (vertical depth 615 feet, average depth on the line of reef, 897 feet) was reached in February of the present year. The Main Reef, including 3 feet of Leader, showed a width of 16 feet. The Leader assayed 16 dwts., and the Main Reef 5 to 6 dwts. per ton.

The average value of the ore crushed (200,785 tons) during the past year was 10.936 dwts. per ton, of which 8.888 dwts. per ton were recovered, namely:—

6.316 dwts. per ton in the mill.
0.469 ,, ,, from concentrates.
2.103 ,, ,, ,, tailings.
8.888

leaving

0.054 dwts. per ton in the concentrate residues.
0.902 ,, ,, ,, cyanide residues.
1.092 ,, ,, ,, slimes.
2.048

The grade of the ore crushed has gradually declined owing to the greatly increased stamping power, the same causes operating as in the case of the City and Suburban Company. The gradual decrease of grade is shown by the average mill returns for the past five years.

	Dwts.	grs.
3rd year of working	11	14
4th ,, ,,	10	14
5th ,, ,,	9	12
6th ,, ,,	8	10
7th ,, ,,	7	17

The Langlaagte Estate.—The reefs worked in this property are Main Reef and South Reef. The Main Reef Leader is not separable from the Main Reef, being close on its hanging wall, and both are stoped together. The average width of the Main Reef (and Leader) is about 7 feet, and the ore tonnage derived from this body is largely in excess of that taken from the South Reef. The Main Reef crushed alone would probably not yield more than 5 dwts. on the plates; but the leavening of it with the rich South Reef brings the grade of the mill-rock up to a little over 6½ dwts., which is probably equivalent to an ore value of 9 or 10 dwts.

The Mines West of the Langlaagte.—Here there is a series of mines in which the Main Reef is unpayable, under present conditions not yielding more than 3½ to 4½ dwts. on the plates. But there is little doubt that some means will soon be found to treat profitably this large body of oxidised ore, having an assay value of 5 or 6 dwts. and a width in places as much as 10 feet. If experiments now being carried out on the direct treatment of oxidised ore with cyanide are carried to a successful issue, large profits will be possible when the upper levels of the Main Reef come to be treated.

The Main Reef Leader, which over most of the ground is separated from the Main Reef by a few feet, is of uncertain value. In places its assay value rises to several ounces, but there are stretches of ground in which it sinks to a few dwts. The uncertain nature of this reef makes it unreliable as a gold producer. The South Reef is the staple producer of this section of the Rand. The gold is chiefly present in the foot wall leader ("South Leader" or "The Leader"), which, though only a few inches in thickness, carries a considerable quantity of gold. Sometimes there are no pebbles in the Leader, which is then represented by a small seam of rich auriferous matter. The South Reef has been extensively worked on the properties of the Main Reef, Unified, Aurora, Aurora West, and the Bantjes; and it is the mainstay of the Durban Roodepoort, United Main Reef, and Princess Mines. There is little doubt that throughout the whole stretch of country lying between the Crœsus and the Banket the South Reef can be profitably worked. Some 24 to 30 inches will have to be mined, of which about 25 per cent must be sorted out as waste. About 8 dwts. will then be obtained on the plates, leaving 4 or 5 dwts. in the tailings for cyanide treatment.

George Goch.—In this mine three reefs are worked, namely, the North, the Main, and the South, each being separated by 20 to 25 feet of sandstone. The North Reef has a stoping width of 3 to 4 feet, and an average ore value of 9 dwts. The Main Reef averages 3 feet 6 inches in the west section of the property, and 5 feet 6 inches in the east section. The grade is also about 9 dwts. The South Reef is stoped to an average width of 2 feet 6 inches, the ore having an average value of 10 to 11 dwts. All three reefs being milled, and some 20 per cent being rejected as waste, the grade of the mill-rock works out at between 10 and 11 dwts. The mill yield during the latter part of 1894 was about 6 dwts.

Simmer and Jack.—The ore mined at the Simmer has an average assay value of a little over 13½ dwts. (fine), and the mill yield is about 8 dwts. In the central section of the mine there is a considerable area or "shoot" of high-grade ore, averaging about 18 dwts. assay value for 4½ feet stoped. The eastern and western sections of the mine, however, are of much lower grade; and it is by the admixture of ore of different quality from various parts of the mine that the grade is maintained at a uniform level. With the new sorting plant lately erected at the present principal hauling shaft, it will be possible to sort out at least 20 per cent of waste, which will permit of larger quantities of low-grade ore being mined without affecting the mill yield. When the proposed 280-stamp mill is at work there is no doubt that the mill grade will be allowed to fall below its present level, as it will then be profitable to mine ore-bodies which under present conditions would yield no profit.

General Conclusions.

In order to gain an idea of the value of the payable ore-bodies of the Main Reef Series, we cannot do better than consult the Chamber of Mines Records of the Output. Thus in the year 1894 we find that thirty-seven companies worked continuously from January to December, and that these thirty-seven companies crushed 2,485,311 tons of ore, yielding 1,182,794 ounces of gold on the plates, and 37,138 ounces by concentration. Further, that the same companies treated 2,150,314 tons of tailings, yielding 486,082 ounces of gold by cyanide treatment. Finally, that the total yield from the thirty-seven companies was 1,715,055 ounces, having a cash value of £5,894,318. In order to analyse these figures it is first necessary to make certain assumptions. First, since many of the companies are still treating old accumulations of tailings in addition to those obtained from the current crushing, it is essential to arrive at an estimate of the ratio obtaining between the current and excess tailings. As a result of our inquiries, we find that on an average about 33 per cent of the rock crushed passes away as slimes, that is to say, that of the total tonnage crushed only two-thirds are collected as sands for the cyanide treatment.

Secondly, a certain proportion of the gold remains in the *residues* or sands that have undergone cyanide treatment. In order to arrive at the amount of this loss, it is necessary to assume an average cyanide extraction; this we find to be about 70 per cent. Finally, in reducing the bullion returns[1] to fine gold, we have had to find an average fineness for each class of gold; the figures adopted by us are 880 for plate gold, 960 for chlorinated gold, and 700 for cyanide gold.

Thus we arrive at the following results, which must not be regarded as scientifically accurate, since there are so many sources of error,[2] but they may be safely taken as near approximations.

Yield per ton of ore crushed :—

	Bullion.	Fine Gold.
	Dwts. per ton.	Dwts. per ton.
Plate yield	9.52	8.38
Concentrates chlorinated	0.14	0.13
Concentrates cyanided	0.16	0.111
Tailings	3.05	2.13
	12.87	10.75
Excess tailings[3]	0.93	0.65
Total yield	13.80	11.40

[1] The Chamber of Mines Returns are made in bullion, the cash value being also given, but on the responsibility of the individual companies.

[2] The Chamber of Mines Returns are compiled from the returns of the companies, and these returns are liable to slight inaccuracies on account of the tonnage crushed being incorrectly estimated, very few of the companies taking the precaution to check their estimates of weighing.

[3] No correction has been made for the fact that the old accumulations of tailings are invariably somewhat richer than those now being produced. The error due to this cause, however, cannot be very great.

The total yield in bullion from the thirty-seven companies during 1894 averaged, therefore, 13.8 dwts. of a cash value of 47s. 6d., per ton of ore crushed.

Carrying the analysis a step farther, we have calculated the ratios of the gold extracted by the different processes in use, the amount remaining in the slimes, and that lost in the residues :—

	Fine Gold.	
	Dwts. per ton.	Percentages.
Caught on plates	8.38	63.7
Extracted from concentrates	0.24	1.8
Extracted from tailings	2.13	16.2
Left in residues	0.91	6.9
Left in slimes	1.50	11.4
Total	13.16	100.0

The original value of the ore crushed works out according to this figuring at 13.16 dwts. of fine gold per ton. The average mill extraction obtained by the thirty-seven companies under consideration was about 64 per cent, and the total extraction averaged 82 per cent.

In the following table we give a list of the thirty-seven companies referred to, arranged in order according to their average plate yield (bullion) for 1894. In the second and third columns we give the highest and the lowest monthly return of the plate yield for the same year; in the third column are the returns for the year 1893; while in the fourth and fifth columns we add the average plate yield for the first two quarters of the present year :—

[TABLE.

AVERAGE MILL GRADE IN DWTS. PER TON.

	1894 Average.	Highest Month in 1894.	Lowest Month in 1894.	1893 Average.	1895 1st 3 Months' Average.	1895 2nd 3 M'nths' Average.
Worcester	22.86	25.00	15.58	14.14	18.39	16.60
Nigel	22.03	27.52	18.06	22.88	17.41	17.02
Ferreira	21.92	26.20	19.13	18.56	21.69	21.38
Robinson	20.56	25.48	17.55	22.11	18.09	19.75
Henry Nourse	16.65	20.48	13.11	17.27	14.74	12.48
Village	12.77	18.15	7.83	10.64	...	...
Champ d'Or	12.64	...	...	6.39	11.69	9.76
Wemmer	12.27	14.68	9.98	16.41	14.74	14.57
New Heriot	11.78	12.84	10.51	13.00	10.87	9.39
Salisbury	11.62	13.49	9.60	15.54	9.83	9.91
Meyer and Charlton	11.32	12.63	9.61	15.98	10.84	13.40
Jubilee	11.28	13.19	10.28	11.34	9.78	9.72
Johannesburg Pioneer	11.00	14.32	9.72	10.30	11.66	12.74
New Rietfontein	10.80	16.00	6.09	23.42	11.41	9.72
Durban Roodepoort	10.03	10.20	10.13	9.63	9.25	9.39
Wolhuter	10.00	10.64	9.16	9.64	10.87	9.06
New Chimes	9.91	11.19	8.38	8.62	10.93	9.54
Stanhope	9.37	10.83	7.73	9.46	8.80	8.06
Princess Estate	9.24	11.88	7.79	6.11	9.65	9.32
United Main	9.14	10.07	8.04	7.66	10.09	11.12
City and Suburban	9.07	14.06	7.57	15.17	8.87	7.08
Jumpers	8.83	10.41	7.96	8.65	8.64	8.86
New Primrose	8.70	10.12	7.67	8.13	7.01	7.91
Randfontein	8.57	11.81	5.22	8.53	10.41	10.03
Van Ryn Estate	8.22	9.58	6.47	6.68	9.88	8.99
Simmer and Jack	8.08	8.64	7.35	7.49	8.31	9.85
Glencairn	7.93	9.28	6.03	8.17	7.79	8.23
Crown Reef	7.80	8.87	7.16	8.74	7.73	8.34
Treasury	7.65	11.66	4.03	12.20	...	...
Geldenhuis Estate	7.51	9.12	6.63	11.42	...	7.15
Metropolitan	7.16	7.88	6.36	4.81	5.94	6.19
May Consolidated	7.12	8.17	6.45	8.27	7.79	7.03
Langlaagte Estate	6.67	8.24	5.37	5.90	8.01	8.32
,, Royal	6.06	8.23	4.47	8.14	4.17	3.44
George Goch	5.88	7.00	4.80	6.92	5.36	6.66
Langlaagte Block B.	5.64	7.40	4.94	6.12	5.36	4.85
Orion	5.41	6.83	4.59	4.57	4.75	4.01
Champ d'Or Deep Level	...	...	...	...	5.79	...
Geldenhuis Main Reef	...	...	...	...	8.42	8.75
Ginsberg	...	...	...	...	13.05	11.00
Meyer and Leeb	...	...	...	...	8.08	11.58
New Kleinfontein	...	...	...	...	7.22	7.06
Paarl Central	...	...	...	...	6.71	7.51

It will be seen from the above table that while the returns of the majority of the companies show for the first half of the present year a slight lowering of the grade of the ore milled as indicated by the plate yield, the following companies maintain their average, viz.: Meyer and Charlton, Crown Reef, Jumpers, George Goch, Glencairn, May Consolidated, Princess, and Chimes; and the following companies have even succeeded in raising their average mill yield:—Wemmer, Pioneer, Simmer and Jack, Langlaagte Estate, United Main, Randfontein, and Van Ryn.

CHAPTER V

DEEP LEVELS—THEIR DEVELOPMENT AND PROSPECTS

FOR the first year or two in the history of the Rand Gold Fields, when, with the exception of one or two properties, it was considered doubtful whether the Main Reef Series would prove generally payable, and the difficulties offered by the hard unoxidised ore to profitable mining and milling appeared insuperable, the enormous potentialities of deep levels were undreamt of, and even the value of ground situated immediately south of the first two or three rows of dip claims was completely ignored. But it was not long before practical men, possessing a serviceable knowledge of mining, and having faith in their own convictions, began to peg off claims on the dip. At first the acquisition of deep level claims was confined to the first two or three rows of claims below the outcrop properties. Gradually, however, as confidence was gained in the possibilities of the banket formation as a gold producer, an increasing amount of the dip ground was taken up.

In January 1890 the May Deep Level shaft struck the reef. At this time claims could still be pegged within 3000 feet of the outcrop, and on the dip of some of the best properties in the Rand, while below the 3000 foot limit the country was practically open to the pegger.[1] It was not, however, till 1891 that any widely extended business in deep level

[1] See the chairman's speech at the annual meeting of the Chamber of Mines, 31st December 1893.

claims properly understood began. By that time the superlative importance of the matter had become evident to the Ecksteins, who are the local representatives of the London firm of Wernher, Beit and Company. This firm, acting with great circumspection and secrecy, began sedulously to acquire blocks of deep level claims on the dip of the best outcrop properties, and thus laid the foundation of the gigantic undertaking known to-day as the Rand Mines, Limited.

The example of the Ecksteins was soon followed by others, notably by the "Consolidated Goldfields of South Africa," who formed the corporation known as the "Goldfields Deep." The price of deep level claims rose rapidly, and the dividing line between valuable gold mining claims and valueless veldt receded farther and farther from the outcrop. To-day there is scarcely a piece of proclaimed ground south of the outcrop of the Main Reef of which the mining rights have not been secured, and claims are even held at as much as 3 miles from the outcrop.

Assuming that the conglomerate beds do not give out in depth, but are continuous over a large area,—a postulate which will, we think, be granted by every one who has studied the geological conditions that prevail in the district, —the problem relating to deep levels turns practically on three questions :—

1. At what depth will the Main Reef Series be found at any given distance from the outcrop?

2. To what extent do the reefs maintain their gold contents in depth ?

3. What is the limit in depth to which profitable mining can be carried?

The first question appears at first sight to be one of comparative simplicity. We know that the whole formation of the Rand dips to the south. We have, therefore,

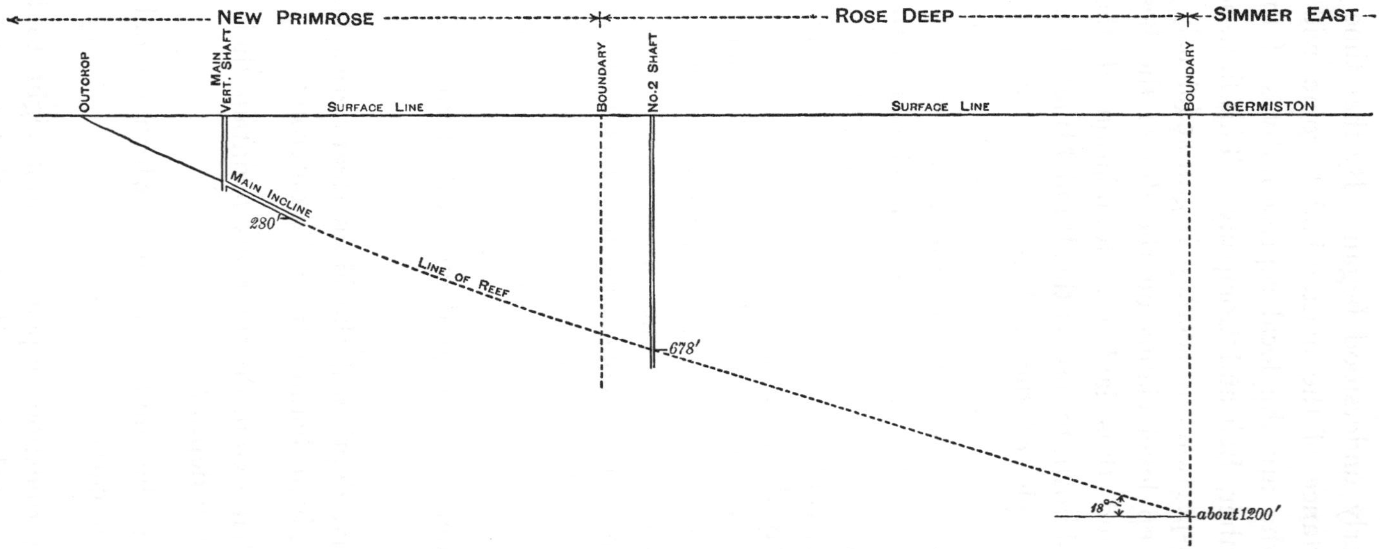

FIG. 46.—Identity of New Primrose and Rose Deep Reefs.

merely to take the angle of slope of the reef-plane, and we can calculate the exact depth at which it will be intersected by a vertical shaft put down at any given point on the dip.

Experience has shown, however, that the problem is not quite so simple as this. The angle of dip observed near the outcrop was found to change as development proceeded downwards, sometimes very gradually, at other times more rapidly. At the present time a considerable amount of evidence has been accumulated by diamond-drill boring, and by sinking deep level shafts; and the general conclusion deducible from the data available is that a very important flattening of the reef does take place in depth. We have prepared a number of cross sections to illustrate this point. (See Figs. 47, 48, 49.)

The angle of dip of the reef at its outcrop varies considerably in different mines, but in most cases it is well over 45°.[1] The dip in the Ferreira, which was 78° in the upper levels, has already decreased to 30° in the lowest part of the mine. In the Jubilee the reef dipped at 75° near the surface; but in the Village, which is immediately south of the Jubilee, it has flattened to 30°. In the Metropolitan the dip has decreased from 70° to 50°. In the upper working of the Henry Nourse the dip of the reef is 80°, while in the deep level shafts, two of which are 500 feet, and one 800 feet from the outcrop, it has fallen to 30°. In the Geldenhuis Estate the dip of the reef is about 35°, while in the shafts of the Geldenhuis Deep it is 28° and 22° respectively. The May Consolidated has a dip of 78° near the surface; in the May Deep the dip is 31°. In the upper workings of the Robinson mine the reef dips at 45°, the

[1] Thus at the Salisbury it is 85°; at the Ferreira, 78°; at the Jubilee, 75°; at the Robinson, 45° to 50°; at the Crown, 50°; at the Metropolitan, 70°; at the Van Ryn, 69°; at the Chimes, 90°; at the Geldenhuis Estate it is only 35°; at the Simmer and Jack, 15° to 20°; at the Primrose, 25°; at the Crœsus, 40°; at the Aurora, 33°; at the Princess, 30°; and at the Banket, 15° to 25°.

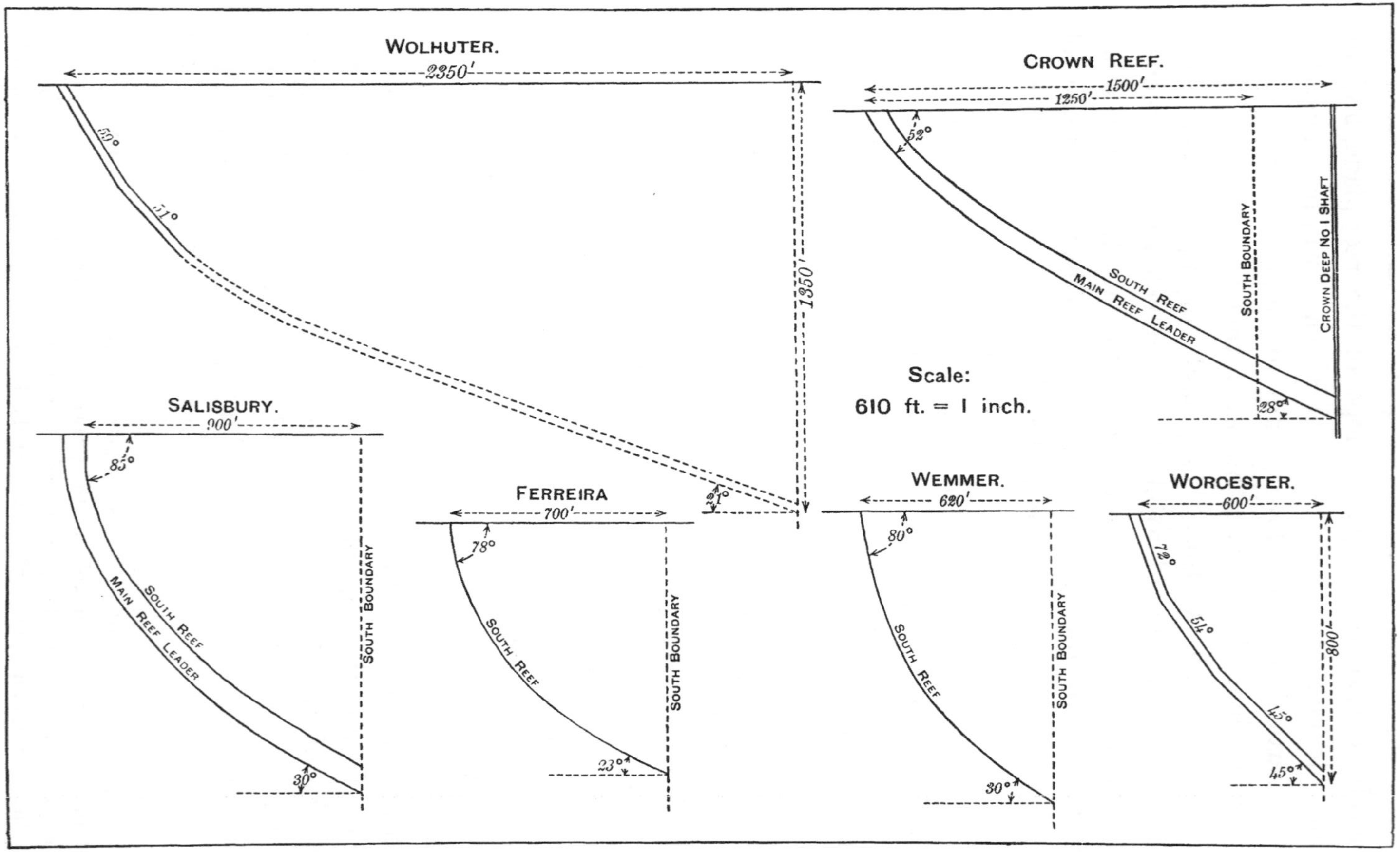

FIG. 47.—Diagrams illustrating the Decrease of the Angle of Dip in Depth.

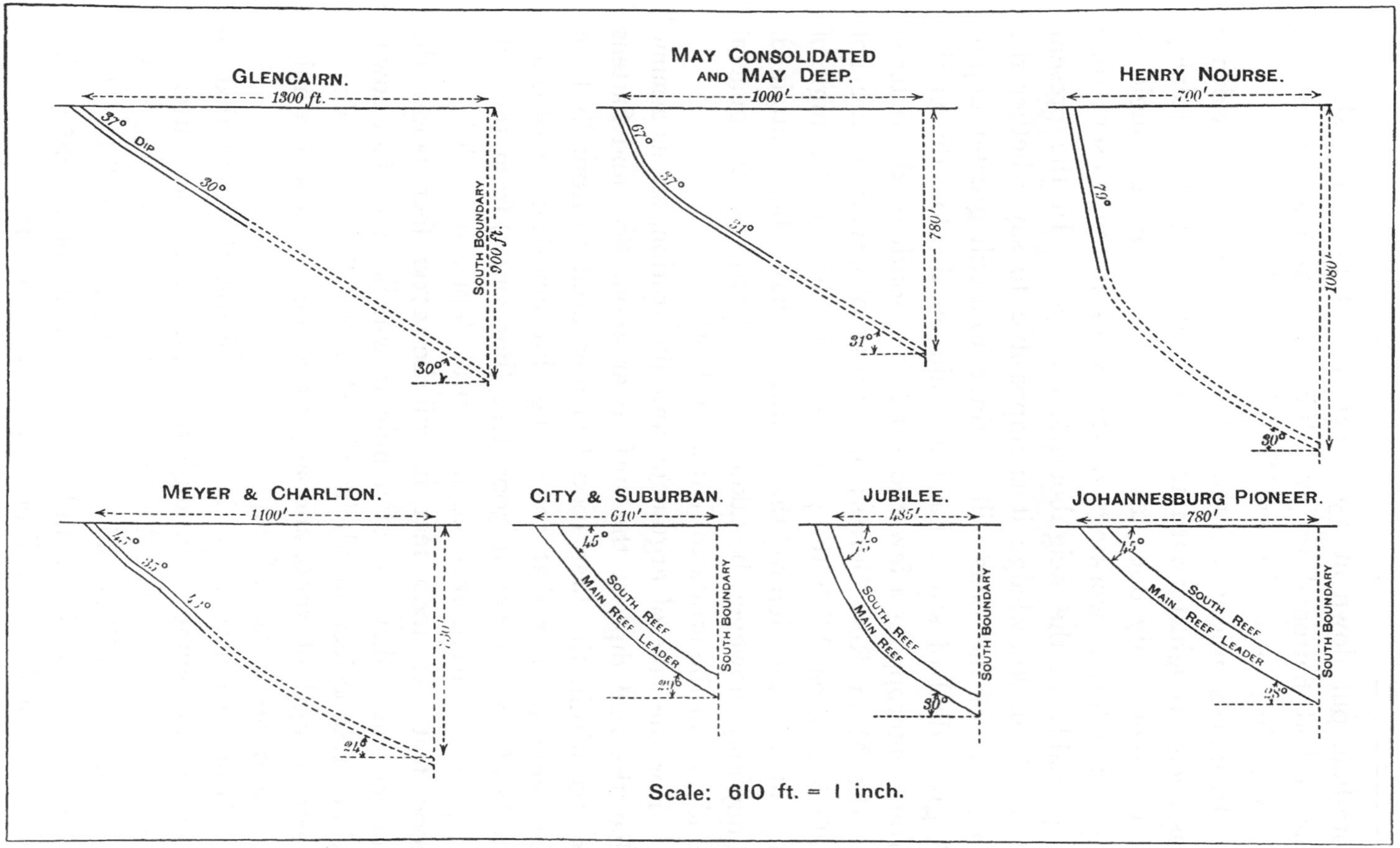

FIG. 48.—Diagrams illustrating the Decrease of the Angle of Dip in Depth—*Continued*.

bore-hole put down at 1500 feet from the outcrop showed that it had flattened to 30°. The reef dips at 52° in the Crown, while in the Crown Deep it is dipping at 27°.

Instances might be multiplied, but sufficient evidence has been brought forward to show that a notable flattening of the reef really does take place. The average angle of dip in the lowest workings of the mines at the present time is probably in the neighbourhood of 30°. In the present state of our knowledge it is impossible to say whether the flattening of the beds will continue to a still greater depth. A great deal of stress has been laid on the low dip of the strata outcropping a few thousand feet south of the outcrop of the Main Reef Series; but these observations are not sufficient to establish the point desired, which is the angle of dip at greater depths than hitherto have been attained. They may, however, be taken as an indication of the general tendency of the strata to flatten in depth.

For the sake of argument and illustration, let us assume that the reef dips at the surface at about 80°, and flattens to 30° within the first 1000 feet measured horizontally from the outcrop, after that continuing downwards at a constant angle of 30°. Now, at 3000 feet (horizontal) from the outcrop it will have reached a vertical depth of a little over 2000 feet; at 4000 feet it will be 2700 feet below the surface; at a distance of a mile it will be 3500 feet down; while at 8000 feet, or about 1½ miles, it will have attained a vertical depth of 5000, and at 15,000 feet, or nearly 3 miles, a total depth of 9000 feet.

If now we assume, as is more probably the case, that instead of continuing at an angle of 30°, a still further flattening, say to 25°, takes place, then at 5000 feet from the outcrop the vertical depth will be only 2800; at 8000 feet it will be a little more than 4000 feet; and at 15,000 feet, or nearly 3 miles from the outcrop,[1] it will be a little over 7000 feet.

[1] To take a specific case, the Rand Victoria bore-hole, which was put down

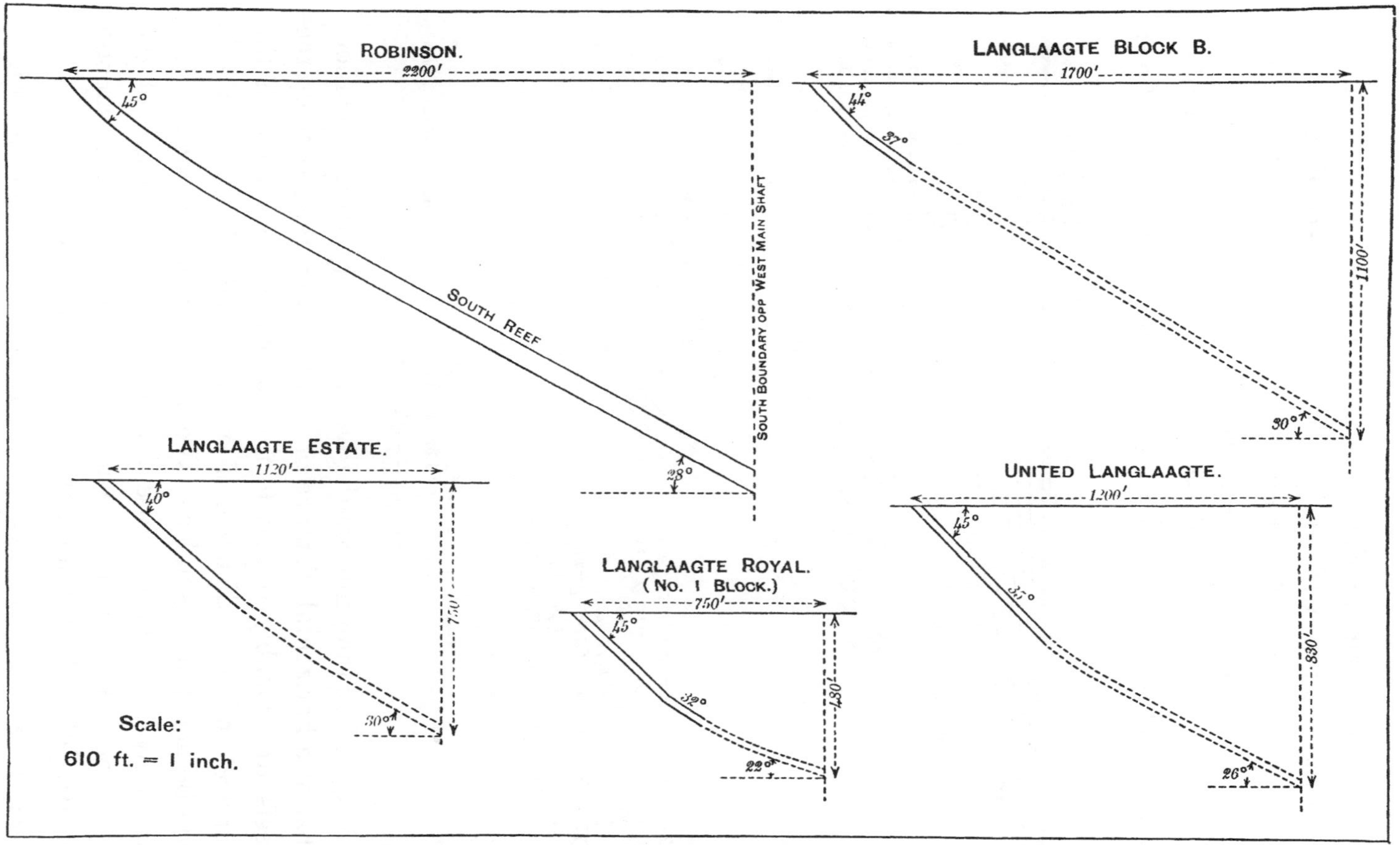

FIG. 49.—Diagrams illustrating the Decrease of the Angle of Dip in Depth—*Continued.*

For ascertaining the vertical depth at a given distance from the outcrop, with varying angles of dip, see Fig. 50.

It will be noticed that our estimates of the depths of the Main Reef Series at given distances from the outcrop are somewhat lower than those arrived at by Mr. Hamilton Smith, and published in his letter to the *Times* of February

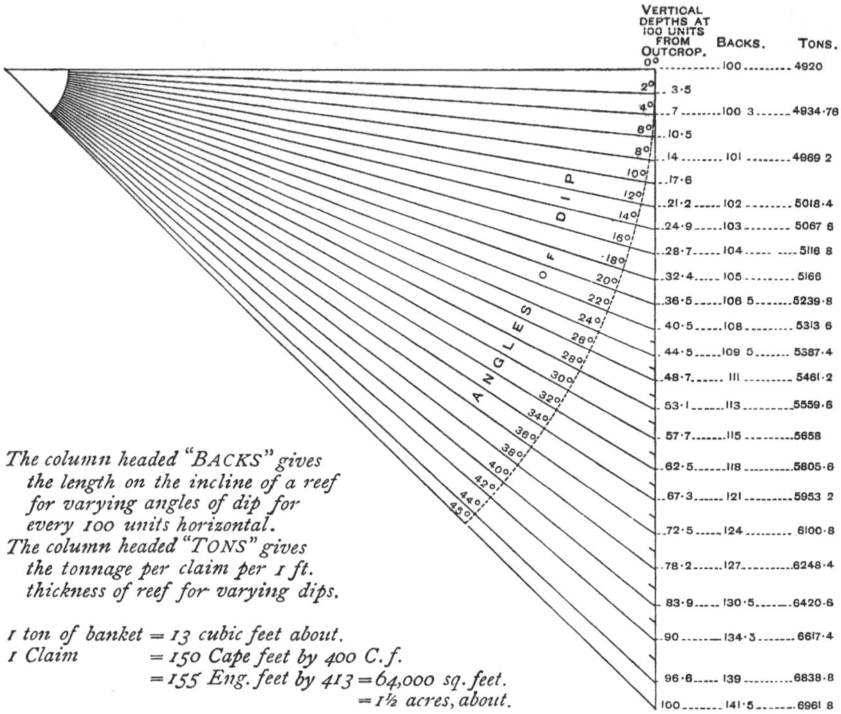

FIG. 50.—Angles of Dip.

1895. Mr. Hamilton Smith gives it as his general conclusion, that at a horizontal distance of 3 miles from their outcrop the reefs are probably 10,000 feet, or about 2 miles beneath the surface, and at a distance of 2 miles their depth as a rule will be not quite $1\frac{1}{2}$ miles. We submit that the facts before us justify an opinion more favourable to the future of the industry. And in this it will be seen we have the support of Mr. John Hays Hammond (see our preface).

south of the Simmer and Jack, at a distance of 4100 feet from the outcrop, intersected the Main Reef at a vertical depth of 2400 feet.

With regard to the second question, namely, *that relating to the maintenance of the ore value in the depth*, we have first the evidence derived from the progress of development in the outcrop properties, and, secondly, the results obtained by boring and sinking deep level shafts. The deepest workings of the outcrop mines have proved the values of the reefs to a distance of 800 to 900 feet on the slope of the reef, and although this distance is very small compared to the greatest depth to which the reefs will probably be worked in the future, it is satisfactory to learn that in regard to aggregate ore value there is no falling off in grade. It is true that there are local fluctuations, but if these seem to indicate a diminished ore value in one mine, in another they tend the other way, so that on the whole the average remains unchanged. Again, it is a common experience for one level to develop a poorer ore than the average of the mine, but the next level below in the majority of cases restores the balance by its improved quality.

The facts on which we have to found our ideas of the value of the ore in the deep level properties consist of a few assays of samples taken from deep level workings and boreholes; and it is evident to every engineer that it is useless to attempt any precise estimates of ore value on the basis of these results. At the same time it must be admitted that the facts hitherto accumulated warrant to a certain degree the confidence that has been displayed on the part of the public in the future of deep level mining, inasmuch as they do not indicate any falling off in quality when compared with the outcrop properties. The available evidence is, however, extremely small, and most of it refers to sampling done at points which, comparatively speaking, are at no great distance from the outcrop. It still remains to be proved that the rich sections of the Rand are not of the nature of large patches rather than shoots, in which case it might be assumed that the variation in quality in deep level properties along the Rand will not necessarily coincide with those in the outcrop properties—in short, that the high-grade ore in the outcrop properties may possibly be replaced by low-grade rock in the deep levels, and the low-grade ore by richer rock. While conclusive evidence in favour of this or any other view is lacking, the assumption that has been made that the value of deep levels is necessarily proportional to the value of the outcrop properties, though perhaps natural, is not exactly justified.

We take the following details from published statements, but of course do not accept any responsibility for the accuracy of the figures:[1]—

A bore-hole put down on the Umbilo Block S. of the May Deep Level 1080 feet from the outcrop yielded the following results:—

Depth at which Reef was struck.	Width of Reef.	Assay Value.		
Feet.	Inches.	oz.	dwt.	grs.
754	20	1	18	0
760	11	0	10	18
777	21	2	9	9

Bore-hole on the Henry Nourse Deep, put down 333 feet from the outcrop:—

Depth.	Width.	Assay Value.		
Feet.	Inches.	oz.	dwt.	grs.
605	30	6	2	4
A few feet lower	30	1	1	0

Rand Deep Level bore-hole, south of the Crown Reef, 1200 feet from the outcrop:—

	Depth.	Width.	Value.		
	Feet.	Inches.	oz.	dwt.	grs.
South Reef	828	18	0	11	10
Main Reef Leader	895	18	13	15	0

[1] Several of these particulars were collected and incorporated in a speech delivered by the chairman at the annual meeting of the Rand Mines Company, Limited, held on December 31st, 1893. They were stated to be given on the authority of the companies concerned.

HENRY NOURSE AND DEEP LEVEL.

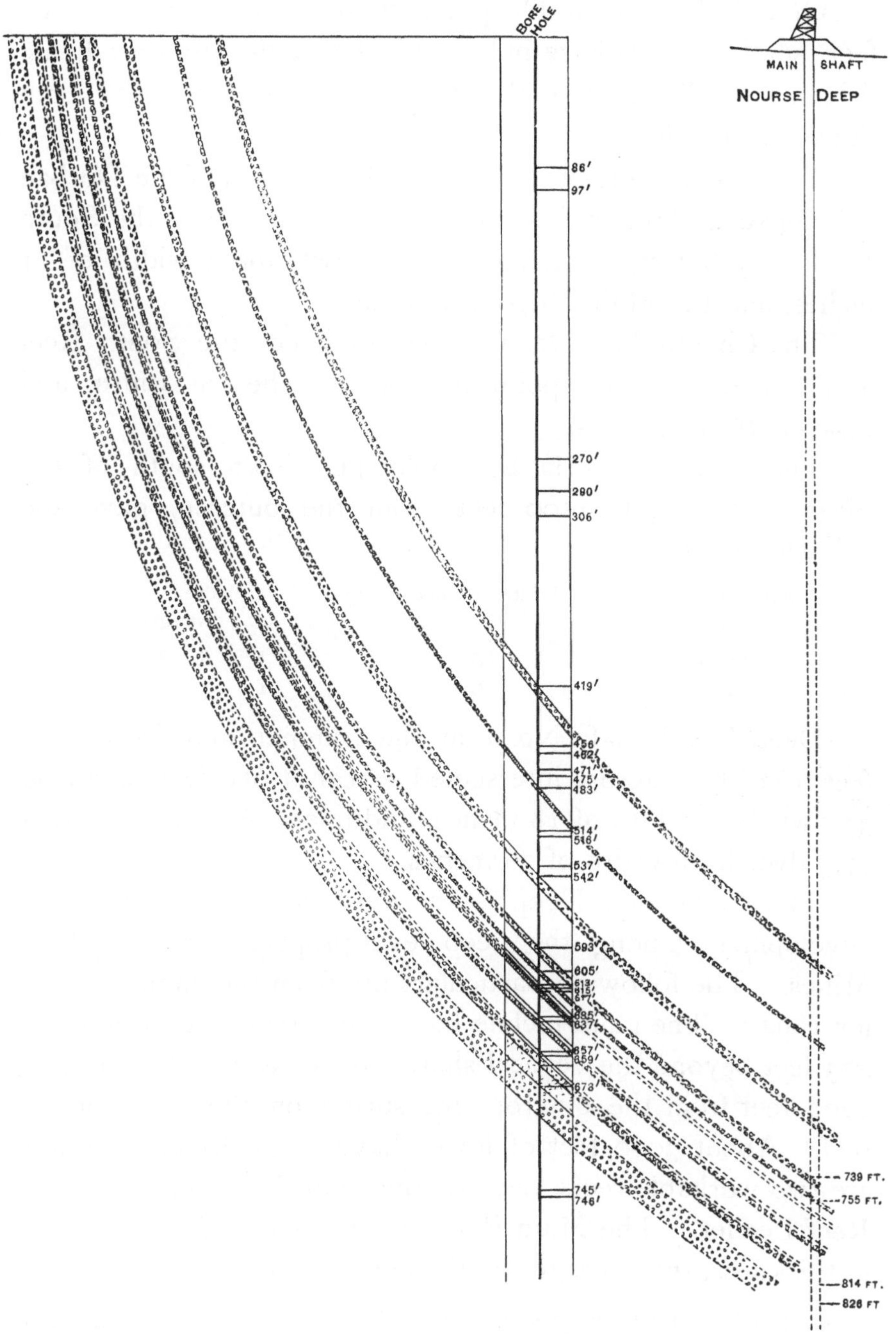

FIG. 51.

One of the Crown Deep shafts has been put down 200 feet south of this bore-hole, and the reefs are now being driven on. The results of sampling substantiate in the main the above results.

The Roodepoort bore-hole, put down south of the Durban Roodepoort, 886 feet from the outcrop, intersected the South Reef at 598 feet, assaying 2 oz. 3 dwt. for a width of 11 inches, and the Main Reef at 620 feet.

The Champ d'Or Deep main shaft struck the South Reef at 410 feet. At this point it was 13 inches in width, and assayed 16 dwt. 2 grs.

The Rand Victoria bore-hole, put down south of the Simmer and Jack 4100 feet from the outcrop, gave the following results :—

Reef Struck at. Feet.	Width in Inches.	Value. oz.	dwt.	grs.
2343	13	1	3	20
2391	9	1	3	20

Deep Level shaft No. 1 at the Simmer and Jack, 2000 feet from the outcrop, intersected a reef at 680 feet, assaying 35 dwt. for a width of 18 inches, and one at 687 feet assaying 17.5 dwt. for a width of 26 inches.

The Geldenhuis Deep is the most advanced in point of development among the deep level properties of the Rand Mines. The following particulars are from the annual report for 1894 :—The main incline has been extended to a depth of 473 feet beyond the vertical shaft (No. 1), at which point it is 2200 feet from the outcrop, measuring on the slope of the reef. About 3000 feet of levels have been driven.[1] There are two workable reefs, namely, the Main Reef and the Main Reef Leader. The Main Reef varies from 3 feet to 8 feet in thickness, the average width lying between 5 and 6 feet.

[1] According to later information six levels are now in progress, the total amount of drifting amounting at the present time to about 10,000 feet. The mill will probably be running at the end of September or beginning of October of this year.

The Main Reef Leader is separated from the Main Reef by a band of sandstone 3 to 6 feet in thickness. The Leader varies from an inch to 3 feet, occasionally occurring in two bands. The development drives being run under the foot-wall of the Main Reef, there have been few opportunities for sampling the reefs, and the results obtained cannot be taken as determinative of the actual qualities of the reefs. 72 samples taken from the Main Reef yielded an average of 9 dwts. 19 samples taken from the Leader averaged 6½ oz. for an average in width of 8 inches.

The Rose Deep west shaft, which struck the reef at a vertical depth of 670 feet, has proved the identity of these reefs with those in the Primrose, and the average value is approximately the same.

The main shaft of the Bonanza Company, whose property is situated on the dip of the Pioneer, Mint, and part of the Robinson, struck the South Reef at 614 feet. Where intersected the reef is 4 feet 9 inches in width, and has a mean average value of 1 oz. 14 dwt.[1]

To what limit in depth can profitable mining be carried? The limiting factors are—

1. Increased temperature.
2. Excessive initial expenditure for equipment and development.
3. Increase of working costs.

1. The well-known fact that there is an increase in temperature proportionate to depth attained, is generally considered the main factor in the causes limiting the depth at which it will pay to mine. This increase in temperature is due principally to the secular heat of the earth, but is also largely dependent on a variety of other causes. Thus the high temperature of mine-workings in carbonaceous or pyritic shales is much increased by the heat developed during the

[1] *South African Mining Journal*, March 16th, p. 498, 1895.

decomposition of the contained mineral matter; again, the proximity of partially cooled volcanic rocks, or the percolation of warm springs, may also exercise a considerable influence. Fortunately for the Rand these subsidiary causes, if they exist at all, are of small account, and the rise in temperature with increasing depth must be ascribed almost totally to secular causes.

In spite, however, of the importance of the question, but few experiments have been made to accurately gauge the exact rate of increase. Mr. Hamilton Smith, during his visit to the Rand in 1894, made some determinations of the temperature of the water in the Rand Victoria bore-hole (2500 feet deep). We quote from his letter to the *Times*:[1] "This bore-hole is in comparative dry ground, so that only a trifle of water flows from it. The water in it is therefore practically quiescent, and represents accurately the temperature of the surrounding walls of rock. These determinations show a temperature of 67.2° Fahrenheit at a depth of 200 feet, increasing in a regular manner to 95.3 at a depth of 2494 feet; this indicates a temperature of 100° at a depth of 3000 feet. Supposing the rocks to have this degree of heat, when a mine is opened up the current of cooler air from the surface, passing through the workings, will reduce this temperature from 5° to 10°; such a heat will add somewhat to the mining costs, but not very greatly at a depth of 3800 feet; the high temperature will probably cause a serious addition to the mining costs." These results indicate an increase of 1° Fahrenheit for every 82 feet.

Since the publication of this report, however, Mr. Hamilton Smith has written to the *Times* to state that the rise in temperature is not so great as that shown by the figures given by him, owing to the inaccuracies in the thermometers used. He proposes to repeat the experiments with more perfect appliances.

[1] *Times*, London, February 19th, 1895.

In a letter published in the *Johannesburg Star* (May 10th, 1895), Mr. Andrew F. Crosse gives the following temperatures taken by him in a shaft on the Ferreira, and in shaft No. 1 of the Crown Deep.

FERREIRA SHAFT.

Depth.	Temperature in degrees Fahrenheit.
123	62.6
221	63.4
317	63.6
409	64.6
496	64.8
572	65.1
638	65.7
698	67.1
758	69.0
808	69.1

CROWN DEEP SHAFT.

Depth.	Temperature in degrees Fahrenheit.
825	66.7
906	69.5
1030	70.7

The increase of temperature indicated by these returns is not sufficiently regular to permit of an average rate being calculated, but the variation would appear to be somewhat less than 1° for every 100 feet, which would mean a temperature of about 89° at 3000 feet, and 108° at 4000 feet, a constant rate of increase being assumed.

It is to be hoped that an effort will be made to obtain an accurate determination of the temperature at the bottom of the bore-hole which is now being put down south of the Meyer and Charlton at a distance of 6000 feet from the outcrop, and which is expected to intersect the reef at a depth of about 3500 feet. Not only is this a matter of scientific interest, but it is of great commercial importance as affecting the future of the industry.

At present we can only be guided in this matter of

temperature by the results of experience in other countries. An interesting letter[1] touching on this subject appeared in one of the Johannesburg papers, from which we extract the following :—

Several inquiries have been made into the question of the possible depth of working mines, notably that of the English Royal Commissioners in Coal, who published a report of their proceedings in 1871. Questions were issued to the managers of coal mines, and evidence was given by those best qualified to do so. The general consensus of opinion, supported by observation in coal mines, seemed to be that the temperature of the strata increases with depth, and that on lengthened exposure to a cooling ventilating current it materially lessens. The temperature of the coal on discovery at Rosebridge Colliery, then the deepest in England, at 2442 feet, was stated by the management to be 93° F., and that it subsequently fell to 63° F.

The mines mentioned below have been worked at the depths given. An idea of what has been done in deep mining can be gathered from this somewhat incomplete table :—

Country.	Mine.	District.	Mineral Worked.	Depth in Feet.	Temperature.	
					Strata.	Air.
1. Belgium	Simon	Lambert	Coal	3489	78° F.	...
2. Austria	Adalbert	Prizebram	Silver, Lead	3279	...	...
3. Prussia	Sanson	St. Andre	Silver	2532	...	...
4. England	Rosebridge	Wigan	Coal	2442	93° F.	85° F.*
5. America	Comstock Lode	...	Gold and Silver	1000	100° F.	87° F.
6. Do.	Do.	...	Do.	2000	130° F.	100° F.

* At bottom of up-cast.

The high temperature of the shallowest mines of the above—those of the Comstock Lode—are due, according to Church, chiefly to the heat given in the kaolinisation of the anhydrous aluminic silicate of the felspathic and amphibolic rocks, an action which also accounts for the heat experienced in Mexican mines in similar ground—extraordinary ones to be neglected in computing the increased temperature for depth in dissimilar strata, which does not fall in exposure.

To the above list the following can be added: Ashton Moss, near Manchester, 2820 feet; Pendleton, near Man-

[1] Signed "Spectator," Wigan, England, 15th May 1895.

chester, 2770 feet; Calumet and Hecla Mine, Michigan, 4426 (to be carried to 5000 feet); and Tamarack Copper Mining Company, Michigan, 3700 feet deep (ultimately to be carried to 6000 feet).

In some parts of Western America the heat at 3000 ft. depth is almost unbearable, while at the Calumet and Hecla shaft there is only a rise of temperature of 4° Fahrenheit in a depth of 4400 feet, although no artificial ventilation is resorted to.

There is no doubt that the adoption of the electric light and the more extended use of machine drills, operated by compressed air, will tend to reduce the temperature of underground workings.

To sum up—the information we have at our command at present is altogether too fragmentary to permit of a final opinion being given as to the possible interference of temperature with profitable mining at deeper levels on the Rand; but we fail to see the slightest cause for uneasiness on this score, the evidence indicating that these fields are at least as well off in this respect as the most favourably situated district in any other part of the world.

2. The question of increased capital expenditure for equipment and development becomes serious when very deep levels are considered. This increased capital tends to diminish profits, for there is a loss of interest on a large sum of money during the period while the mine is non-producing, the redemption of which must be allowed for before the mine-owner can count his profits. As an example, let us consider what initial expenditure is necessary to equip and develop a fair-sized deep-level mining property to supply a 200-stamp mill. For the Geldenhuis Deep, which in reality can scarcely be termed a deep-level proposition at all, the reefs being intersected by the two shafts at a depth less than 1000 feet, it is estimated that a working capital of £350,000 will suffice to provide for everything. Of this probably £150,000 will

be required for the 200-stamp mill, cyanide works, etc., leaving £200,000 for the equipment and development of the mine and for interest on the capital.

Let us now take the case of a property situated at such a distance from the outcrop that the shafts will have to be sunk to a depth of 3000 feet before the reefs are intersected. In a case of this kind 5-compartment shafts would be necessary, one compartment being reserved for pumping, and the remaining four for two pair of balanced skips. Probably 12-inch pumps would be required; at all events it would be necessary to provide adequately against possible water difficulties. The total expenditure would amount roughly to £650,000, allocated as follows:—

Sinking and equipping two shafts to a depth of 3000 feet .	£300,000
Pumps (two sets), say	100,000 [1]
Mine development	100,000
Mill, cyanide works, etc..	150,000
	£650,000

Supposing this expenditure to be distributed over a period of six years, the interest on the capital would amount to another £100,000, bringing the total working capital up to £750,000.

Let us assume that the property consists of 200 claims of a nominal annual value of £500 each,[2] and estimate what such a property might be expected to return on the actual capital expended. With an average dip of 30° and $5\frac{1}{2}$ feet of reefs worked, each claim may be expected to yield 27,000 tons of ore. Two hundred claims will, on this calculation, contain 5,400,000 tons.

The profit per ton in working Rand ores will no doubt be

[1] This item is purposely put at a high figure in order to allow for possible contingencies.

[2] This assumption is entirely arbitrary, the actual value varying according to the situation of the ground and the amount of risk the purchaser is prepared to take.

somewhat increased by the reduction of working costs,[1] and it has been estimated that it will soon average 20s. per ton in the outcrop mines.[2] On the other hand, at a depth of 3000 feet working expenses will be but slightly increased by the extra cost of hauling, pumping, and ventilation. We are surely on the safe side, therefore, in assuming an average profit of 15s. per ton.

5,400,000 tons of ore yielding 15s. per ton profit would give a total profit of £4,050,000. Against this we have—

Working capital	£750,000
Claim value at £500 . . .	100,000
Total capital . . .	£850,000

Reckoned on this sum the total yield works out at 476 per cent. A 200-stamp mill will crush 350,000 tons per annum, and would therefore exhaust the property in sixteen years.

The total yield distributed over this period is equivalent to an annual dividend of 30 per cent. Allowing 6 per cent for the redemption of the capital, we arrive at a net return on the outlay of 24 per cent per annum for a period of sixteen years.

These estimates may be quite beside the mark. We cannot expect to approximate very closely to the truth on account of the variable nature of the factors.

The purchase value of deep level claims, as already pointed out, depends on the amount of risk that the purchaser is satisfied to take, and this risk grows with the distance from the proved ground near the outcrop of the reef. Thus we can fix a line at such a distance from the outcrop that the value in the light of possible dividends is reduced to *nil*. Between this line of no value, and the line of maximum value as represented by the outcrop properties, it is possible

[1] See page 271.

[2] See *Rand Mines First Annual Report* (1893

SCHEDULE OF DEEP

	Property.	No. of Claims.	Position in relation to Outcrop Property.
Predominant Interest held by the Rand Mines.	Rose Deep . .	133	Deep Level of Primrose . . . . .
	Geldenhuis Deep .	211	Deep Level of Geldenhuis Estate and Geldenhuis Main
	Jumpers Deep . .	224	Deep Level of Jumpers and E. Section of Heriot
	Nourse Deep . .	269	Deep Level of Henry Nourse, Ruby, and W. Section of Heriot
	Crown Deep . .	191	Deep Level of Crown Reef and Robinson .
	Langlaagte Deep .	300	Deep Level of Langlaagte Estate and Langlaagte Royal
	Durban Roodepoort Deep	230	Deep Level of Durban Roodepoort and Roodepoort Deep
	Ferreira Deep (Rand Exploring Syndicate)	142	Deep Level of Worcester, Ferreira, and Wemmer
	South Rand . .	110	Deep Level of Crown Deep . . . .
	Rand Mines Claims .	...	Deep Level of Goch, Metropolitan, Wolhuter, Meyer and Charlton
Predominant Interest held by the Consolidated Gold Fields.	Knight's Deep . .	181	Deep Level of Witwatersrand
	Glen Deep . .	183	Deep Level of May and Glencairn . . .
	South Simmer (included at present in Simmer and Jack)	...	Deep Level of Simmer and Jack and part of Primrose
	Simmer East . .	458½	Deep Level of Rose Deep
	Simmer West . .	250	Deep Level of Geldenhuis Deep
	South Salmond . .	420	Deep Level of Glen Deep and Knight's Deep
	Village Main . .	92	Deep Level of Wemmer, Salisbury, Jubilee, and W. part of City and Suburban
	Robinson Deep . .	127 and Mynpacht	Deep Level of Crown Deep, Robinson, and Rand Exploring Syndicate
	Roodepoort Deep .	135½	Deep Level of United Main Reef . . .
	Champ d'Or Deep .	188	...
	Nigel Deep . .	659	Deep Level of Nigel Mynpacht . . .
	Central Nigel . .	546	Deep Level of Nigel Mynpacht . . .
	Bonanza . . .	11	Deep Level of Pioneer . . . . .
	Rietfontein Deep .	85	Deep Level of Rietfontein Estate, East Section
	South Rietfontein .	101	Deep Level of West and Central Sections .
	Lancaster . . .	...	Deep Level of South of the York (late Emma)

LEVEL PROPERTIES.

No. of Vertical Shafts.	Depth at which the Reef has been or is expected to be struck on the Shafts.		Remarks.
	Actual.	Estimated.	Expected to Crush.
	Feet.	Feet.	
2	(1) 868, (2) 670	...	Latter end of 1896
2	(1) 583, (2) 830	...	100 Stamps at the end of Sept. 1895
2	...	(1) 750	Latter end of 1897
2	(1) 985	(2) 1200-1400, (3) uncertain	Latter end of 1896
2	(1) 1030, (2) 1057	...	Latter end of 1896
2	...	(1) 1000	Latter end of 1897
2	...	(1) 1500, (2) 1300	
...	...	...	Work not started
...	...	...	Work not started
...	...	...	Work not started
...	...	...	Rand Mines are part owners
7	(1) 680, (3) 355	(2) 1100 (4) Cyanide shaft, 350 (5) Rand Victoria, 2500 (6) Rudd shaft, 2500 (7) Rhodes shaft, 2500	
2	...	(2) 1000	
2	...	(1) 2300, (2) 1900	
1	{ South Reef at 598 Main Reef at 660		
2	(1) 453, (2) Incline 780		
5	...	500-800	
3	...	Nearest at 1550	
1	...	(1) 690	First Quarter 1896
1	...	700	
...	...	...	Amalgamated with New Rietfontein Estate

to construct a curve showing the variation in units of value. Such a curve must be constructed so as to show a rapid increase in value at first, as we pass from depths at which physical causes may prevent profitable working to depths where working costs are likely to be normal; then comes a long straight portion of the curve, where the increase in value is quite gradual; finally, another upward sweep as we approach the outcrop mines, where the value is absolutely a known quantity.

Development in most of the deep level properties is being pushed energetically forward, those most advanced at the present time being the Geldenhuis Deep,[1] Rose Deep, South Simmer, Nourse Deep, Village Main, Bonanza, Crown Deep, Langlaagte Deep, Champ d'Or Deep, and Roodepoort Deep. Work is also progressing in the Jumper's Deep, Robinson Deep, and Durban Roodepoort Deep. New properties are constantly being formed by the amalgamation and adjustment of the interests of neighbouring claim-holders. Among the most recent consolidations thus effected are those of the Glen Deep, Knight's Deep, Simmer East, Simmer West, Nigel Deep, Central Nigel, Western Nigel, Sub-Nigel. The controlling powers in these properties is almost entirely vested in the hands of either the Rand Mines, Limited, or the Consolidated Goldfields of South Africa.

The preceding table gives the deep level properties in existence at the present date. It shows the size (number of claims) and position of each property in regard to the outcrop property, of which it is the deep level. The number of vertical shafts is also given, and the depth to which they have been or have to be sunk to fetch the reef.

[1] The Geldenhuis Deep is expected to start milling in September or October of this year, and the Rose Deep, Nourse Deep, and Crown Deep towards the latter end of next year.—*Rand Mines Report* for 1894, p. 17.

CHAPTER VI

MINING PRACTICE ON THE RAND

SINCE the banket beds seldom appear at the surface continuously for any great distance, a certain amount of prospecting work is always necessary in following up reef extensions in the direction of their strike. The more highly mineralised beds readily decompose under the influence of the atmosphere, and little is seen of them at the surface, whereas conglomerates carrying little pyrites are mostly hard and durable. The Main Reef Series, which embraces the most valuable reefs in the district, has seldom been found with a conspicuous outcrop.

The nature and amount of prospecting that may be necessary depends on the geological indications that exist at the surface, and on the depth of the superficial deposits overlying the banket formation. The uncertainty as to the result of prospecting for extensions of reefs in the absence of definite surface indications arises from faulting or from variations in the strike or other irregularities. After a preliminary examination of the surface for any indications, such as outcropping parallel beds of sandstone, shale, or conglomerate, it is usual to begin by trenching across the direction in which the reef is supposed to be striking. If trenching prove useless owing to the great thickness of overlying deposits, then a shaft is sunk from which cross-cut drives are made, or two or three shafts are sunk and connected in solid formation by cross-cuts.

Where big faults exist it is often a matter of great difficulty to locate the broken portion of a reef, and in many cases it has been found expedient to resort to boring by diamond-drill. During the last six or seven years a great deal of boring has been carried on throughout the district, both with the object of locating extensions of reefs, and of proving the depths of reefs on the dip side of their known outcrops. Much valuable information has been obtained in this way. While, however, boreholes are of the greatest use in locating beds, they are unsatisfactory as regards any information they afford as to the reef values, owing to the great variations in reef thickness and gold contents, and to the smallness of the core-sample obtained.

Laying Out and Developing.

Under the head of *developing* we propose to treat briefly of the general practice on the Rand in the laying out of mines, and in such operations as shaft-sinking, driving cross-cuts and levels, and making winze connections, which, in the systematic working of a mine, must precede the actual winning or stoping of the ore.

Since the early days of the fields, when the industry was in its infancy, great strides have naturally been made in all directions, and among other things that have been impressed on managements, often as the result of adverse experiences, is the importance of developing adequate reserves of ore. After the general recognition of the permanence of the reefs followed by an increase in the magnitude of the various mining undertakings, including large milling plants, it was realised that not only a continued monthly expenditure in developing operations, but also a large initial outlay in permanent works and preliminary development, had to be met. Thus while five years ago most of the working mines lived pretty much from hand to mouth, there are few companies to-day that are not run on a sound basis with regard to the development of adequate ore reserves.

The general plan on which the mines are laid out is that adopted in vein-mining. A property is opened by one or more working shafts, according to its length along the strike. These shafts are either vertical, when sunk to strike the reefs on their underlay, or inclined when they follow the inclination of the beds. At intervals, varying with the dip and the backs it is deemed advisable to open up, cross-cuts are driven to intersect the various reefs, and drives are carried east and west along the reefs; further connections are made between the different levels by winzes. The ore bodies to be worked are thus blocked out in rectangular areas, that can be attacked from any or all sides, according to the methods of stoping adopted, and the demand for ore. This general scheme, which is of course familiar to all who understand the first principles of mining, was preceded on the Witwatersrand by much more primitive methods. In the early days, when owners considered it waste of time to sink shafts, or lacked the necessary confidence, quantities of ore were taken out along the outcrop from long open cuttings by means of windlasses thrown across the top. However, it was not long before shafts and the general working plan above sketched out came into vogue. The regularity usually maintained by the banket beds renders it possible to lay out works with comparative certainty, and to make close estimates of the time and outlay involved.

Shafts.—It has been stated that one or more main working shafts are required according to the length of the property. In the case of outcrop properties less than 1500 or 2000 feet in length (*i.e.* along the strike of the reefs) generally only one shaft is equipped and carried down as a main working shaft, but for larger properties two main shafts are considered necessary in order to facilitate development and ensure adequate ventilation. In deep level properties, unless very small, it is advisable to have two or more shafts. The two shafts at the Crown Deep are about 1700 feet apart, the three at the Nourse Deep about 1000 feet, and the two at the Geldenhuis Deep 1300 feet apart. The total length of the Crown Deep property is nearly 6000 feet, and it is probable that a third shaft will be sunk. The Nourse Deep and Geldenhuis Deep properties are respectively 6400 feet and 4000 feet in length. On the Amalgamated Simmer and Jack properties there are at present in use and in progress ten shafts for purposes of development and ore-raising. Of these one is a main incline in the eastern section of the property, the rest are vertical shafts, and three of these are laid out to intersect the reef at a depth of 2500 feet. The scheme which is being followed in exploiting the western and main section of the Simmer and Jack differs somewhat from that usually adopted. The property is a large one, and the western section, which is divided by a fault from the rest of the ground, is over 4000 feet in length. The inclination of the beds from the surface is low, so that vertical shafts sunk at considerable distances from the outcrop cut the reef at comparatively small depths. The shaft system may be described as consisting of three pairs of vertical shafts,[1] a line connecting the two shafts of each pair being approximately in the direction of dip. The average distance between the shafts in the direction of strike is about 1500 feet. The shafts of each pair will be connected by main winzes from 700 to 1000 feet in length, which will be continued below the deeper shafts. The winzes will be used as self-acting inclined planes for conveying the ore to the lower vertical shafts, in which all the raising of rock coming from above will be effected. One or two levels below the point where the deeper verticals intersect the reef will be developed from these verticals. Below the deeper verticals the main winzes will be used as haulage inclines. The method of working

[1] Excluding the three deep level shafts.

properties by means of vertical shafts sunk near the dip boundary, while more correct from an engineering point of view, and perhaps likely to prove more economical in the long run, has not been generally adopted on the Rand. Since the depth to which it would be necessary to sink in order to catch the reefs would be very much increased if such a plan were adopted, greater time would elapse and greater initial outlay would be necessary before a profit-earning stage could be arrived at, and these are drawbacks which few shareholders care to face.

Up to within the last few years it was the general practice in laying out the mines to sink vertical shafts comparatively near the outcrop, the point selected varying with the dip of the beds as far as it was known. Since in many places the dip exceeded 60°, while the degree of flattening in depth was not known, there was at that time much to be said in favour of vertical shafts. It is true that cross-cuts of varying length had to be made at each level, but the sinking necessary to reach a given depth was less, and there were the usual advantages of vertical shafts, viz. smoothness in running and minimum expenditure in maintenance. Had the high angles of dip, in many cases as much as 70° and 80°, continued to great depths, as it was at first supposed they would, it is probable that vertical shafts would still be much more common in outcrop properties than they are, but it soon became recognised that there is general flattening of the beds, the most rapid decrease being in those cases where the beds are most highly inclined at the surface. It is now practically proved that at a vertical depth of 800 feet the angle of dip seldom exceeds 32°, and it is also noteworthy that at that depth there is a near approach to uniformity of dip from point to point along the Rand. The recognition of these facts has led gradually to an almost entire revolution in the matter of shaft location and sinking in outcrop mines; so that while in 1891 probably not half a dozen of such mines in the district were worked by main inclined shafts, those that are not so worked to-day are exceptions to the general rule. There can be no doubt that the great saving of time as well as cost, resulting from the adoption of inclined shafts, is sufficient justification for the change. In the majority of outcrop properties now working the shafts are inclined from surface downwards, the original vertical shafts having been abandoned, but in some instances shafts carried down vertically for several hundred feet have, after intersecting the Main or South Reef, been continued on the underlay. In deep level properties vertical shafts are of course inevitable, but they are almost invariably sunk near the north or rise boundary, and are turned off on the underlay when the reefs are struck.

With three exceptions, viz. the main shafts at the Langlaagte Royal, Unified, and Primrose, which are circular,[1] all the vertical shafts on the Rand are rectangular in section, varying in dimensions within timbers between 11 feet by 5 feet and 26 feet by 6 feet. They are divided

[1] These round shafts were put down by Mr. L. Hamilton, who considers them preferable to rectangular shafts on the score of greater cheapness (both in sinking and timbering) and strength, and as affording better means of ventilation. The following details are supplied by Mr. Hamilton :—

New Primrose—
Diameter, 11 ft.
Depth, 400 ft.
Excavation area, 95 square ft.
Compartments, 2. (One for Cornish pumps, etc. One large compartment with one double and one single cage. Large cage with 40-lb. rail guides on pump dividers. Small cage with three rope guides.
This shaft is equal to a 3-compartment rectangular shaft.

Langlaagte Royal—
Diameter, 15 ft.
Depth, 500 ft. (to go to 900 ft.)
Excavation area, 176 square ft.
Compartments, 4 and 8 spaces for pipes, etc. *a a*, Pressure pipes, Moore's hydraulic pump ; *b*, rising main, do. ; *c*, spare rising main ; *d*, steam-pipe to stand-by pump ; *e*, compressed air main ; *ffff*, spare spaces for pipes, ropes, etc.

PLATE II.—Headgear, Sorting-House, Mill-House, Tailings Wheel, and Collecting Vats, New Primrose Mine.

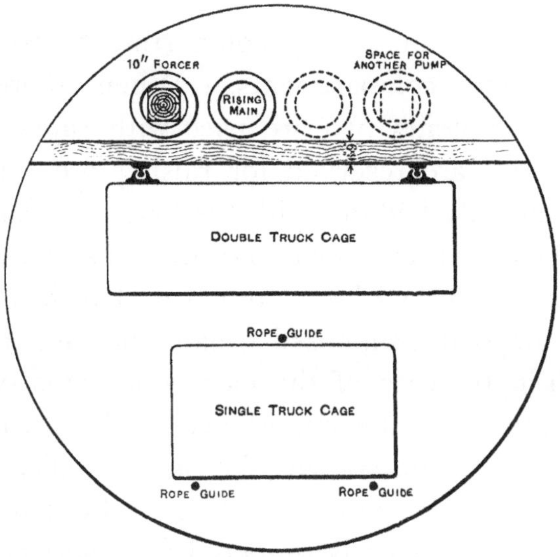

Fig. 52.—Section of Round Shaft at the New Primrose.

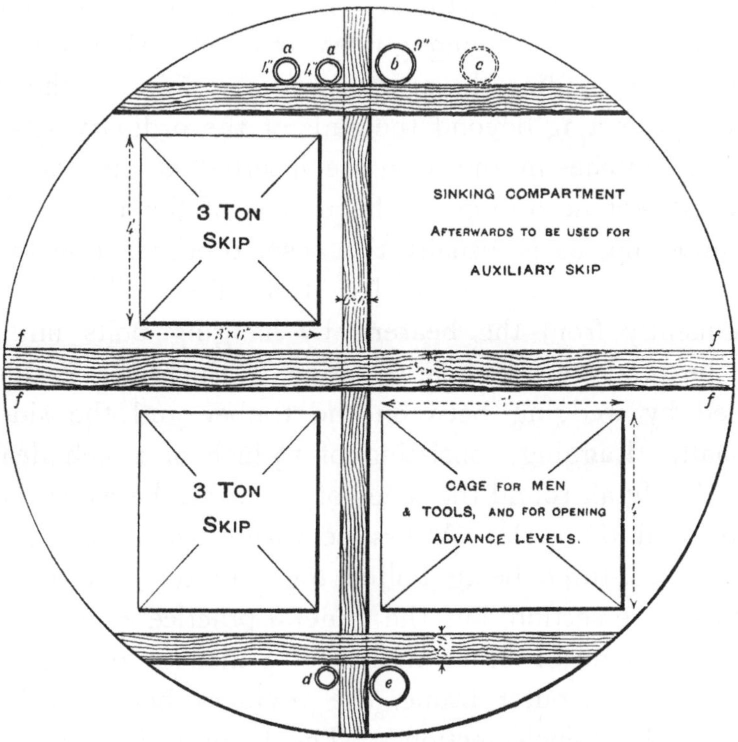

Fig. 53.—Section of Round Shaft at the Langlaagte Royal.

into a ladder-way and pump compartment and two, three, or four hoisting compartments. When there are four hoisting compartments two are used with cages for general traffic, while two are reserved for raising ore. Rectangular shafts are generally laid out with the long side parallel to the strike of the formation; some, however, have been sunk at right angles to the strike, a practice that has nothing to commend it beyond a supposed greater firmness of the shaft walls, which in the case of the close quartzites of the Main Reef formation is of little importance. The great drawback in such shafts is, of course, the impossibility of turning them off on the underlay to advantage. Permanent vertical shafts are invariably closely timbered from surface to depths, varying from 100 to 200 feet. The timbering consists of square sets, from 4 to 6 feet apart, with corner posts or studdles. The compartments are separated by dividers which are used for fixing guides and rails, while they also assist in withstanding the pressure on the sides of the shaft. Bearers projecting beyond the ends of the ordinary sets, and resting in hitches in the rock, are inserted at intervals from 50 to 100 feet according to the nature of the rock. When the timbering, as is usually the case, proceeds downwards, successive sets are suspended from the collar set and subsequently from the bearers, by hanging bolts until the next bearers are fixed in position. Each set is further secured by wedging between the timber and the sides of the shaft. Lagging, consisting of 1½-inch or 2-inch deals, is fitted closely all round the sides of the shaft between the set timbers and the rock. At first sets were usually made up of 3-inch deals, three being bolted together to form timber of 9-inch square section, but the general practice now is to use solid pitch pine or other timber of 8-inch or 9-inch square section for the outer frame, the dividers being of 8-inch or 9-inch by 6-inch section.[1] The lagging is seldom con-

[1] In most of the deep level shafts Australian Karri wood is being used exclusively.

tinued below a depth of 200 feet, the rock at that depth being undecomposed and very firm ; sets are, however, often continued down to much greater depths, and dividing timbers can never be omitted since they are essential to the carrying on of work in the shaft. According to the Government Mining Regulations it is compulsory to divide ladder ways by bratticing from hoisting compartments.

In the deep level properties controlled by the Rand Mines Ld., most of the shafts are 20 or 22 feet by 6 feet in the clear, and have three hoisting compartments, 6 feet by 4 feet 3 inches, or 4 feet 6 inches, and a pump and ladder compartment 6 feet by 6 feet or 6 feet 6 inches. Such shafts are designed for large outputs from depths of 1000 to 2000 feet. The Simmer and Jack deep level shafts laid out to intersect the reef at 2500 feet are 26 by 6 feet in the clear, and have four hoisting compartments.

Inclined shafts starting from the surface either follow one or other of the reefs all the way, or, as is frequently the case, they are located to the north of the outcrop, and are sunk at such an angle as to intersect or approach the reefs at a considerable depth. The object of the latter course is to avoid, as much as possible, curves and changes of inclination in sinking. Figs. 54 and 55 illustrate the relations of inclined shafts to the reefs in the cases of the Robinson West Incline and the City and Suburban Central Shaft respectively. The latter shaft commences about 85 feet to the north of the outcrop of the South Reef, which at the surface dips at 45° south. The shaft was sunk at an angle of 30°, the intention having been to cut the South Reef twice. The reef was intersected between the second and third levels, after which the shaft was continued in the overlying beds. Had the anticipated degree of flattening been found the reef would have again been intersected before now, but as the outcrop dip has continued with rather more than usual persistence, and as, moreover, the reef has been dropped by a fault, the

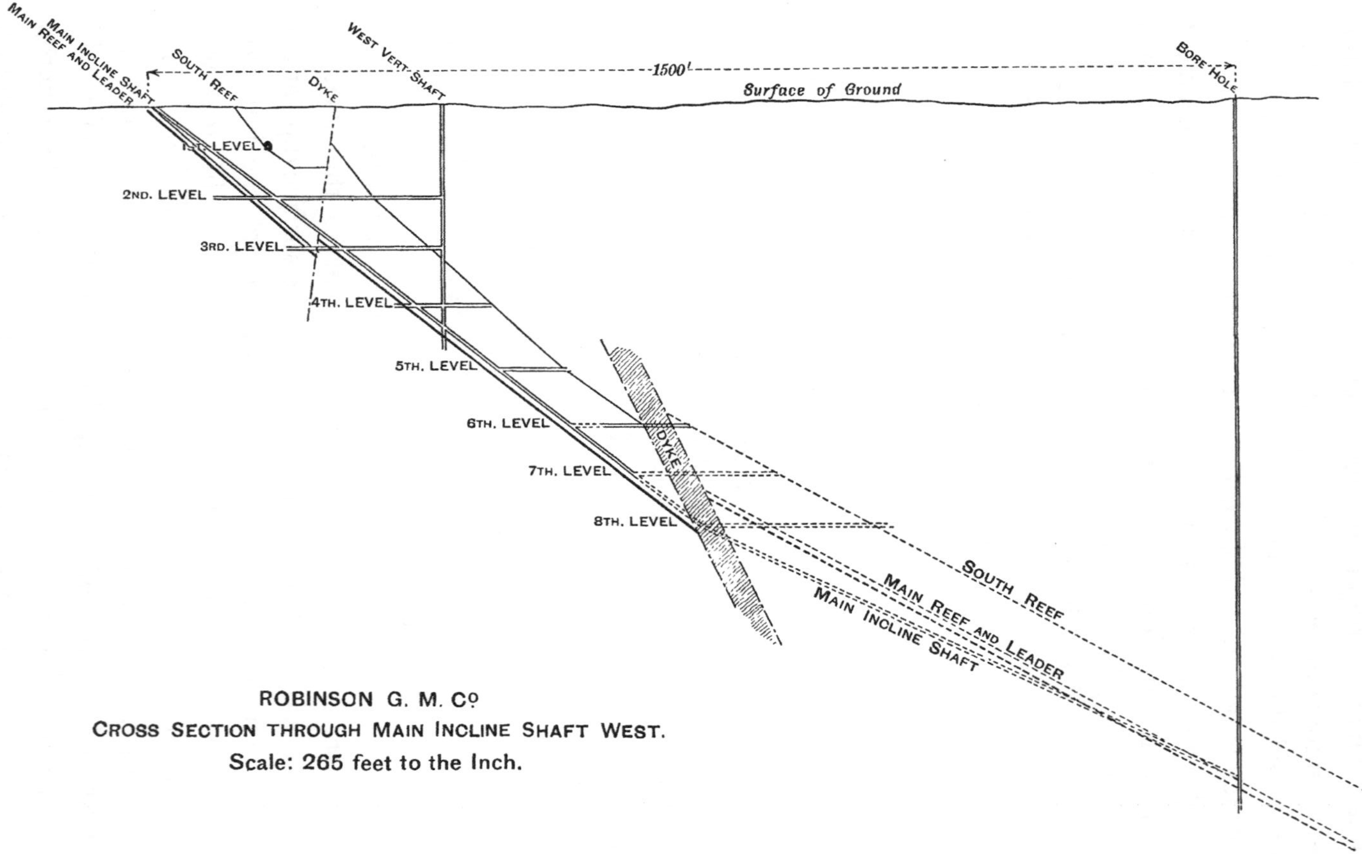

ROBINSON G. M. Co.
CROSS SECTION THROUGH MAIN INCLINE SHAFT WEST.
Scale: 265 feet to the Inch.

FIG. 54.

Photographed by Davies Bros.

PLATE III.—HEADGEAR (VERTICAL SHAFT) AND ENGINE AND BOILER-HOUSE, SALISBURY MINE.

second intersection has not yet been accomplished. The dimensions of inclined shafts vary similarly to vertical shafts. Some of the largest, such as the City and Suburban abovementioned, are 20 feet in length. They are divided into two or three skip-ways, a pump compartment, and ladder way. The timbering near the surface is in complete sets about 5 feet apart, and the footwall for the first 20 or 30 feet is often built in with masonry. Side lagging is seldom necessary after the first 50 or 100 feet, but the hanging wall is of course closely timbered for several hundred feet, and sills and props dividing the compartments are continued throughout.

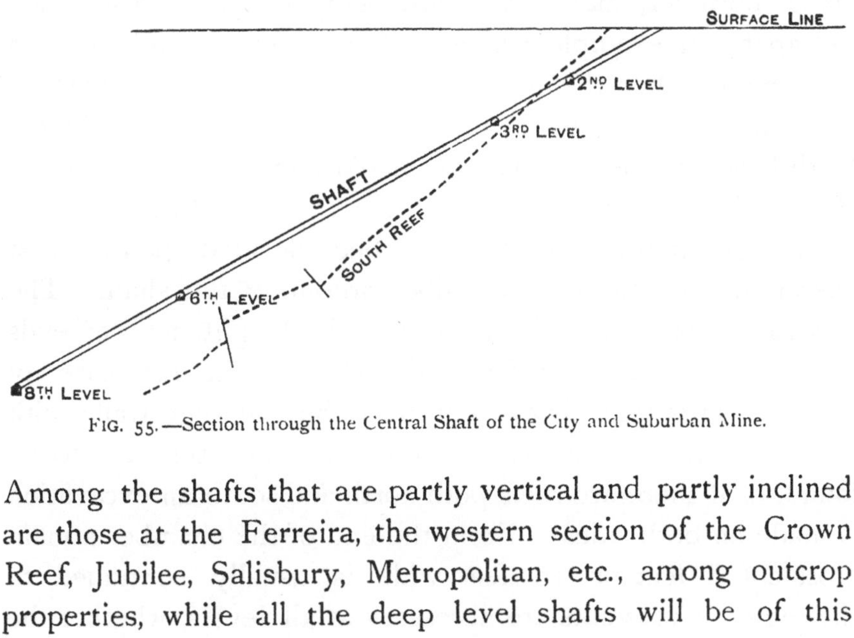

FIG. 55.—Section through the Central Shaft of the City and Suburban Mine.

Among the shafts that are partly vertical and partly inclined are those at the Ferreira, the western section of the Crown Reef, Jubilee, Salisbury, Metropolitan, etc., among outcrop properties, while all the deep level shafts will be of this class.

Levels.—In the early days the first levels were driven at as little as 30 to 50 feet from the surface, followed by a second at about 100 feet, but of late years the tendency has been to increase the interval within practicable limits. The expense of driving levels is so great that in the interests of economy it is clearly important to keep them as far apart as possible. The practical considerations that put a limit on the amount of backs that can be economically developed by one level are, the difficulty, expense, and time involved in raising and sinking long winzes, and in handling the ore in large stopes. In most of the outcrop mines the reefs at the levels now worked are dipping at angles which are too low to allow of the unaided movement of broken rock down the stopes, and consequently a great deal of shovelling is required in order to remove the ore. It is chiefly owing to the time and labour that would otherwise be expended in this operation that the backs or interval between successive levels on the slope of the reef seldom exceeds 130 feet. At the Geldenhuis Deep, however, main levels are being driven 200 and 300 feet apart, an example which will doubtless be followed in other mines when the problem involved, viz. the economical removal of ore from the stopes, receives practical solution. With levels 300 feet apart intermediate levels can be driven afterwards if necessary, so as to give 150 feet of backs.

The position of a level having been decided upon the first operation is to cut out a chamber adjoining the shaft. The excavation is usually 8 or 10 feet in height, and extends perhaps 15 feet away from the shaft. In the case of inclined shafts it is always cut into the hanging wall; with vertical shafts it is on that side on which the reefs are to be found, or if there are reefs both north and south an excavation is made on both sides. From the end of the chamber a main cross-cut 8 feet or 10 feet wide is usually continued to the reefs. Cross-cuts are necessary at all levels, whether the shaft be inclined or vertical, since with rare exceptions two or more of the reefs are workable; but in the case of inclines the length of the cross-cuts is practically limited by the distance of the reefs from one another, and is more or less constant at all levels. If the shaft be inclined and sunk on the upper reef, as occasionally happens, there is of course no cross-cut from the end of the chamber. The drives are then started in the hanging of the reef two or three yards away from the shaft, and are turned back towards the reef as shown in Fig. 56 *aa*. Any reefs underlying are subsequently opened up by cross-cuts north from the main drive on the upper reef (see Fig. 56 *bb*). The drives on the various reefs are extended in both directions (east and west) to the boundaries of the property. It is usual in driving to keep the reef in or near the roof.

It frequently happens that two pay reefs are sufficiently close together to be developed by the same drive, as, for instance, the Main Reef and Main Reef Leader, or a group of reefs or "leaders" separated by varying thicknesses of sandstone, such as often constitutes the South Reef. In such cases it is the usual practice in driving to keep the lower seam in the drive, or in the hanging of the drive, the others being exposed only at intervals for purposes of sampling or determining their position.

Drives are generally carried about 5 feet in width by 6 feet 6 inches in height, or somewhat larger where machine drills are used. The grade allowed for purposes of drainage is about ¼ inch per yard, or 1 in 150.

Timbering of levels, except in preparing stopes, is never necessary at depths greater than 150 feet, and consequently in the majority of Main Reef mines is now a thing of the past. Even in developing nearer the surface timbering is only occasionally necessary, as in decomposed dyke matter and very soft sandstone.

Winzes.—Winzes are sunk (or raised as the case may be) at intervals from one level to another to prepare the ground for stoping and to secure ventilation. As in the case of levels the intervals are variable, but are as great as is compatible with the attainment of the above ends. They are seldom less than 200 feet, and are sometimes as much as 600 feet or 800 feet apart.

Winzes are generally sunk as far as the broken rock, and water can be economically dealt with, in order to avoid loss of

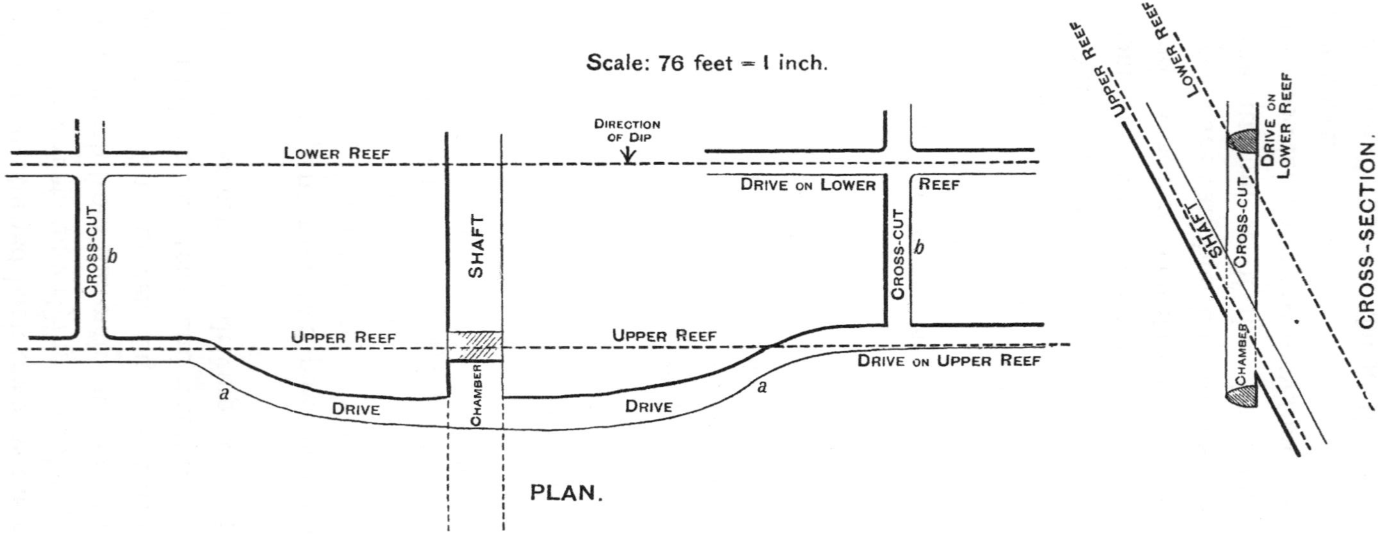

FIG. 56.—Plan of Levels and Cross-cuts.

time in getting ventilation, but a balance of the connection is usually left to be raised from the next level, since raising to a height of 40 or 50 feet is a more expeditious and less costly operation than sinking. Compressed air as well as steam pumps are frequently used in sinking winzes.

Rates of Development.—The rates attained in sinking and driving, of course, vary greatly according to circumstances. Apart from the difference between soft, oxidised "red" rock in the upper levels, and hard, unoxidised "blue" rock at greater depths there are very marked variations in the stratification, jointing, and texture of the beds. Shales are always taken advantage of when they form the footwall of a reef, as in the Blue Sky, Kleinfontein, Van Ryn, Nigel and the Heidelberg Roodepoort, and they are perhaps the easiest kind of ground in depth. On the other hand, a large compact body of conglomerate, such as the Main Reef below oxidation level, is the hardest and toughest rock worked in the district. As affecting rates of sinking, it may be mentioned that the quantity of water met with, especially below the first 200 to 300 feet, is seldom great, *i.e.* within the limits of present experience.

The average monthly rates of eight months' steady sinking at the Rose Deep with machine drills in blue rock were 67 feet and 60 feet in Nos. 1 and 2 shafts respectively. No. 1 shaft is 21 feet 6 inches, and No. 2 shaft 16 feet by 6 feet in the clear. The record for one month was 93 feet. As affecting the averages, it should be mentioned that two pump-chambers were cut in the shaft during the period, and that the shafts were completely timbered. At the Langlaagte Deep in January last No. 2 vertical shaft was sunk 92 feet in red rock by hand labour. At the Robinson Deep, in eleven months ending March 1894, No. 1 shaft was sunk 690 feet, and No. 2, 738 feet from the surface, the averages being 63 feet and 67 feet per month respectively. The depths sunk in March were 84 feet and

78 feet respectively. Only hand labour was employed. The main vertical shaft of the Bonanza Limited was sunk 631 feet in 10 months by hand labour. The shaft is 15 feet by 5 feet in the clear. The average speed for the last 7 months in blue rock was 60 feet per month. The shaft was timbered throughout and a sump was cut below water level. Three shifts were run. In sinking inclined shafts greater speed can be made; as much as 165 feet was sunk in one month at the Salisbury, while 167 feet was sunk in the same time at the New Primrose. These rates are, however, exceptional, the average speed attained in continuous sinking being less.

The rates of driving by machine drills vary from 50 to 150 feet, probably averaging about 80 or 90 feet per month. Higher averages obtain in non-producing mines where only development works are being carried on, than in producing mines where the shafts and drives are concurrently used for the extraction of the ore. The rates of driving attained by hand labour vary from 30 to 70 feet per month, averaging perhaps 40 feet. Although there is no doubt that much greater speed can be attained in driving and sinking inclines when machine drills are used, the speed of sinking vertical shafts on the Rand by means of machine drills has not hitherto been very much greater than that attained by hand labour.

Machine drills are very extensively used. Indeed, there is hardly a mine in the district working in blue ground that has not an air-compressing plant, while in some properties as many as forty drills can be run. Since it is generally admitted that development by machine drills is more costly than by hand labour, their extensive adoption on the Rand must be attributed directly to the impossibility of developing by hand labour at the rate necessary to meet the demands of large milling plants. Large mills, again, are the result of the policy of extracting the total gold in the shortest time practicable, a policy that for deposits of the stability and permanence of the Main Reef Series is, without question, financially sound.

In sinking, two or three drills are used; in an ordinary drive one, and in cross-cuts or drives intended for a double track, or where it is required to make greater speed, two. The drills mostly used are the Eclipse (Ingersoll-Sergeant), the Slugger, Little Giant (Rand Drill Company), the Hirnant, the Banket, and the Climax, the two first being in greatest favour.

Stoping.

In stoping, *i.e.* the breaking down the ore blocked out by drives and winzes, as above described, both the over and underhand methods are adopted. Frequently they are used in conjunction, as illustrated in Figure 57, A.

Formerly it was almost the universal practice to stope overhand, but of late years underhand stoping has been largely adopted for various reasons, of which the principal are:—

1. The expense of preparing overhand stopes is great on account of the scarcity of native timber.
2. The Kaffirs, who are often but little skilled as miners, are very much more efficient in underhand than in overhead striking.
3. Underhand stopes are in many cases more cheaply worked. Mr. J. H. Johns, in a report on the Pioneer Mine, states that, "where reefs are small and hard, underhand stoping is 20 per cent less costly than overhand."
4. It was formerly objected to underhand stoping of reefs, which, owing to small size, necessitated the breaking of much waste rock, that the waste could not be separated in the stopes, but with the sorting arrangements now in vogue this objection is removed, and the economy of underhand stopes may in such cases outweigh the additional cost of hoisting and sorting.

In preparing overhand stopes, the drives are protected either by stull timbering and lagging, or by leaving an arch or strip of the reef over the main drive, and starting the stope from a small back drive, as shown in Fig. 57, B. This latter method is without doubt the more economical in soft rock.

At intervals of from 30 to 50 feet small raises, to be used as ore passes, connect the main drive with the back drive, and before stoping is commenced wooden shoots with rough check-gates are fixed in these passes for convenience in loading trucks. As the stope proceeds the lower worked-out portion may become gradually filled with waste rock, in which case the passes are continued upwards by building loose pack walls on either side.

When the dip of the reef is low, say less than 30°, it is a common practice to carry a stope face forward almost parallel to, but slightly inclined towards the bottom of the winze, as shown in Fig. 57, C. On account of the small angle of dip there is no danger to men in the lower part of the stope. This method, like ordinary underhand stoping, has the advantage of requiring no back drive or timbering at the lower level, it being necessary only to leave a pillar or arch at the bottom, and to put up short passes into the stope as required (Fig. 57, D). Any waste picked out is thrown back in the stope. This method, which is sometimes called "breast stoping," is economical, but if used exclusively it has the disadvantage of limiting the length of the working faces.

As a matter of fact, in most of the mines one finds all three classes of stope adopted.

As already stated, two reefs are sometimes near enough together to be developed by the same drive. They may, however, be too far apart to be worked out in one stope. Some managers under these circumstances take out the upper reef by means of a succession of perpendicular raises,

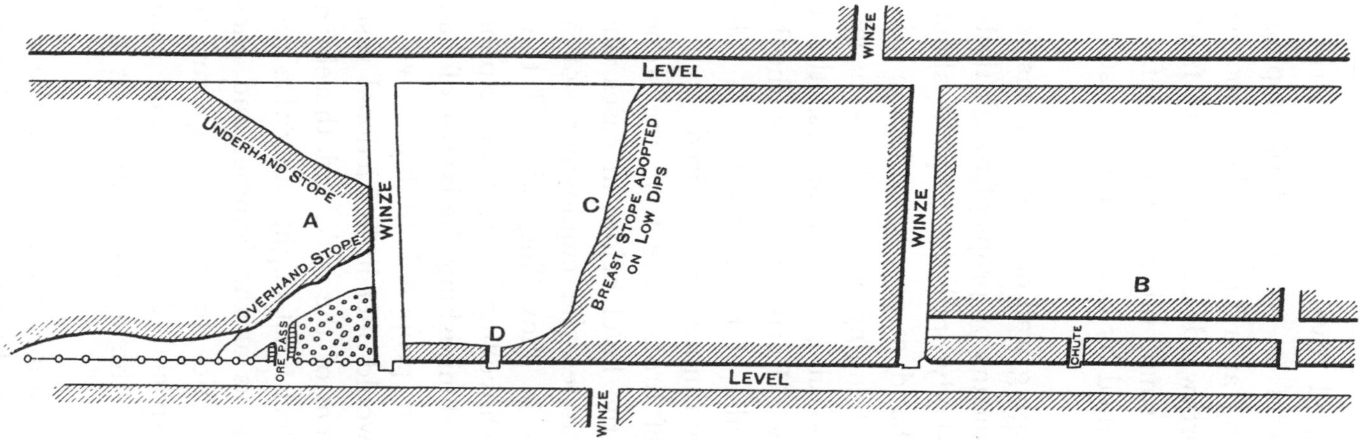

Scale: 66 feet = 1 inch.

FIG. 57.—Diagram showing Methods of Stoping.

afterwards connected over the main drive, and carried upwards on the reef so as to afford stoping faces; or if the beds are dipping at a high angle, a succession of cross-cuts are driven northwards or southwards, from which stopes are opened up by small raises on the reef. Ventilation is in most cases obtained by holing into the stope on the other reef.

Where the hanging wall of a reef is unsafe, protection is secured either by means of props of rough timber, by leaving pillars of the ore-body until the stope is about to be abandoned, or, in overhand stopes, by filling in with waste rock from the foot or hanging wall. It is usual to leave as little timber as possible in old stopes, the cost of good props being heavy. Fortunately, below the level of oxidation, where the walls of the reef are as a rule in excellent condition, very few timbers are required.

The moving of broken ore from the stope faces to the loading-shoots at the level below, in cases where the dip is under 35°, necessitates the employment of a considerable number of Kaffirs in shovelling, while if the dip is in the region of 20° this shovelling becomes an expensive operation. With the object of diminishing the labour of handling rock in stopes, it has been proposed to lay down ore-shoots of sheet-iron, which would offer little frictional resistance to the downward movement of the rock, and thereby enable two or three Kaffirs to effect what might otherwise require a dozen. In very flat slopes, as at the Simmer and Jack and at the Orion, it is the practice to run in temporary tracks, and to use trucks of a special design, but this cannot be done economically where the average dip is at a greater angle than a few degrees.

The width of a stope, except under the most favourable conditions of foot and hanging walls, is always greater than that of the ore-body mined. When a reef is 3 feet, or over, in thickness the conditions of the wall may be such that the

Photographed by Davies Bros.

PLATE IV.—SURFACE WORKS AT THE ORION MINE.

banket can be broken clean, *i.e.* without an appreciable admixture of sandstone or quartzite. This, however, rarely happens; and further, there are as a rule quartzite partings in a reef that must be broken down with the banket. Again, the payable banket seams are frequently too small to be mined alone; thus, for instance, the South Reef and Main Reef Leader are generally of less thickness than it is possible to stope, so that large quantities of valueless quartzite become mixed up with the pay rock. Sometimes as much as 80 per cent of the rock broken in stoping is quartzite or banket, too poor to pay milling expenses. Worthless banket, sometimes unavoidably broken from a poor seam lying close to a pay reef, must go through all the subsequent operations of handling and treatment, since the difference to the eye between payable and unpayable banket is not sufficiently apparent; but quartzite waste can and is separated and sorted out with a very important effect on the grade of rock sent to the batteries. Sorting is to a small extent carried on in the stopes, where the larger lumps of dead rock are thrown aside or used for filling, pack walls, etc. The major part of the sorting, however, is effected on the surface, and with this subject we shall deal in the following chapter.

Where a reef has a soft footwall, a method of "resueing" is sometimes adopted, *i.e.* the footwall is first removed and thrown aside. The reef can then be broken down comparatively free from waste. This method of stoping is generally adopted with shale footwalls, with the additional advantage that the shale or argillaceous matter, which yields excessive slimes in milling, is kept out of the mill-rock.

The native labour employed in hand drilling and in handling rock in stopes varies from $1\frac{1}{4}$ to $2\frac{1}{2}$ Kaffirs per ton of ore. The dynamite expended per ton varies from $\frac{3}{4}$ to $1\frac{1}{4}$ lb. Candles are almost invariably used for lighting working faces in stoping and developing.

Tramming and Raising.

Tracks of about 18-inch gauge are laid down in all working drives, in order to tram the ore to the shaft. In main cross-cuts it is usual to lay a double track. The rails, which weigh from 12 to 16 lbs. per yard, generally rest on light steel sleepers set in the rubble with which the drive is levelled up, but in some mines wooden sleepers are preferred as they make a firmer track. The trucks are usually imported, and are made of steel plate, and are of a capacity ranging from 10 to 16 cubic feet. Sometimes they are built on the mines of $1\frac{1}{2}$-inch deals and light iron. End-tipping or side-tipping trucks are used according to requirements.

Although mules are employed underground in some mines, for instance at the City and Suburban, the labour of tramming generally falls on Kaffirs, one or two being necessary for each truck according to its size and the length of the drive. The trammers fill their own trucks from the ore-shoots or from the working faces of the drives, and take them to the shaft, where they are either run into cages or dumped into skips or into ore bins at the shaft. A white man or a native at each loading station is responsible for the tally of trucks according as they come from one reef or another, or contain waste rock. In order to keep the tally correctly the trammers are generally assigned a particular drive or portion of a drive.

In vertical shafts the rock is hoisted either in cages, into which the trucks are run, or in skips. In inclined shafts it is hauled either in skips or, if the inclination is low throughout, in the mine trucks direct, in which case there are switches between the level and the incline, or flat sheets on which the trucks are turned. Cages are usually built of light steel, with a 3-inch wood floor carrying rails, and a roof and sometimes sides of steel plate. In a few shafts cages are run on rope guides, but by far the most general practice is to fix wooden runners to the dividers on each side of the hoisting compartments. Cages are seldom made to carry more than one truck at a time.

In the more recently equipped vertical shafts self-dumping[1] skips, which are found more expeditious, are used in place of cages; where a shaft is partly vertical and partly inclined it is of course, necessary to use skips. Skips are made of riveted steel plate from ½ inch to 1 inch in thickness, and are generally of the form sketched in Fig. 62.[2] The skip rails in inclined shafts weigh from 20 to 40 lbs. per yard according to the capacity of the skip, which varies from 1 to 4 tons.

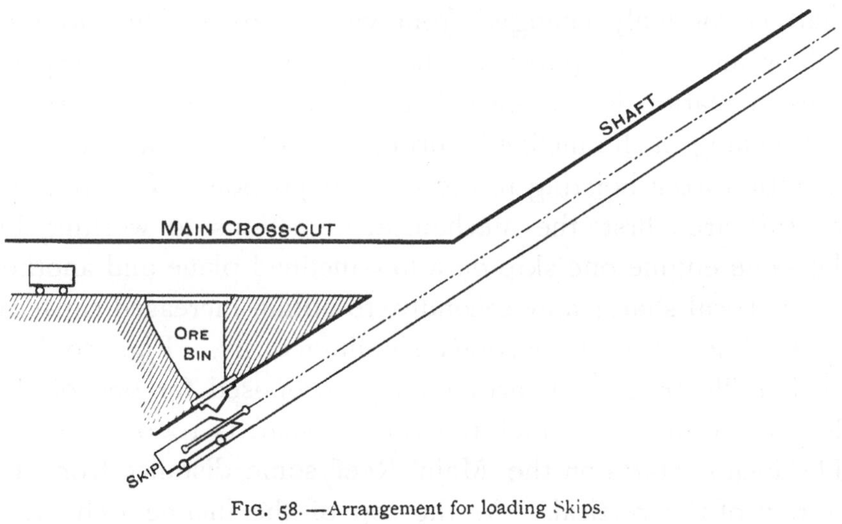

FIG. 58.—Arrangement for loading Skips.

The manner of filling skips formerly most general was to tip the mine trucks directly into them, but in most mines now it is customary to cut ore bins or shoots in the floor of the main cross-cut, opening into the roof of the incline by bin-gates with shoots, through which the skips are filled. The discharge shoot of the bin is generally in the direction of the inclination of the shaft, and if the inclination is low, the top of the skip is partly cut away, so that they can be filled readily,

[1] See surface equipment, p. 141. [2] Surface equipment, p. 142.

see Fig. 58. By this method time is saved and regularity of working secured, for delay of trucks about the loading stations is avoided, since they can be at once tipped into the bin, irrespective of the work being done in the shaft; and the skips again can be filled at any level at any moment, as long as there is rock in the station-bin. Further, the delivery of rock to the sorting floors need be subject to no intermission.

Where shafts are not carried to great depths vertically, and where there is no sudden transition from the vertical to the underlay, the hauling roads are continuous throughout the shaft, as in the Ferreira and the Metropolitan; but where the shaft is suddenly changed from vertical to an inclination of 30° or less, as is generally the case in deep level propositions, a station is established at the bottom of the vertical, and hauling in the inclined portion is carried on as a distinct operation from hoisting in the vertical portion. The reasons for this are: first, the mechanical objections to working by the same engine one skip on a low inclined plane and another in a vertical shaft; and, secondly, to obtain increased capacity by hauling in both portions simultaneously. Fig. 59 is a section illustrating an arrangement proposed for one of the deep level mines in which the vertical shaft is 1000 feet deep. The incline starts on the Main Reef some distance from the bottom of the vertical. At the top of the incline a chamber will be cut out to the south for the accommodation of a compressed air or electric winding engine, and to the north of the incline at the side of the vertical shaft a bin of some 350 tons' capacity will be excavated. The skips running in the inclined shaft will dump their contents into this bin, from which the skips in the vertical shaft can be filled as desired. A at the bottom of the bin is a check-door over a small hopper B, with a gate through which the skips can be filled.

Fig. 60 is a section for which we are indebted to Mr. V. M. Clement, showing an arrangement proposed for one of the

South Simmer and Jack shafts. Here the engines for hauling in the incline will be on the surface, and the hauling ropes will be taken down the vertical and round pulleys at the bottom into the incline. It will be seen that the skips are to dump some 80 feet below the top of the incline into a shoot leading to the bottom of the vertical, where the main loading station of the vertical shaft is situated. There is undoubtedly

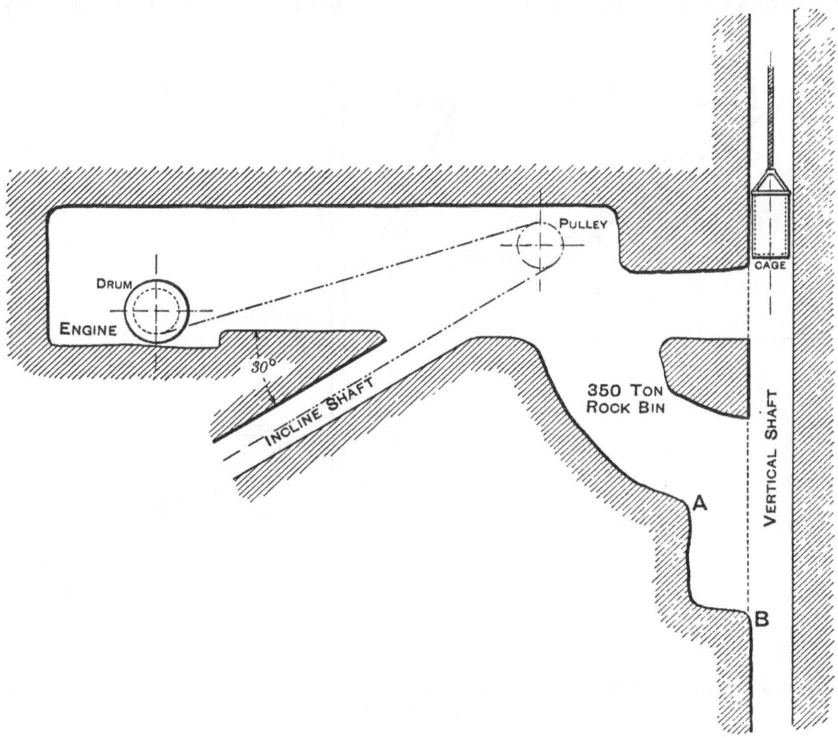

FIG. 59.—Proposed Arrangement for working Deep Level Mines.

economy in having a steam engine on surface, rather than a compressed air or electric engine below. As shown in the sketch, it is proposed to have only one upper loading station which by a system of shoots can serve for three or four levels above it, thus saving the expense of cutting and fitting out a number of stations, and at the same time simplifying the general arrangements.

Underground stations and main cross-cuts are generally lighted by electricity.

MINE DRAINAGE.

The quantity of water met with in the Witwatersrand mines is seldom great, as far as present experience goes, and as a rule it is heaviest near the surface, so that the first pump lift relieves a mine of most of its water. It is only in very

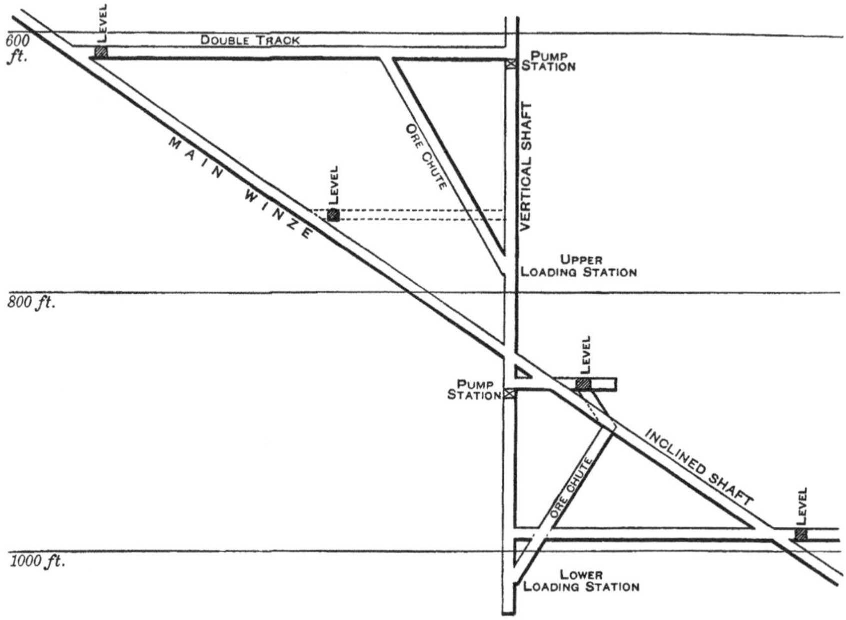

FIG. 60.—Proposed Arrangement for reducing number of Stations at one of the South Simmer and Jack Shafts.

exceptional cases that the quantity of water pumped in a shaft exceeds 150,000 gallons per day, and therefore, although 12-inch plunger lifts have occasionally been brought into requisition, it is seldom that Cornish pumps of more than 9-inch working diameter are required. The top lift of a Cornish pump is generally from 8 inches to 10 inches, followed by other plunger sets from 6 inches to 9 inches diameter, at intervals of from 150 to 250 feet vertical.

In the Crown Deep No. 1 shaft, 1000 feet deep, from

which development is now being carried on, there are no pumps deeper than 250 feet, all the water below this depth being handled very easily by means of skips.

FIG. 61.—Riedler Pump in use at the Geldenhuis Deep. Compound engine driven by compressed air, the pump being placed at the bottom of the vertical shaft at a depth of about 600 feet. Speed at this head 150 revolutions per minute. Plungers $3\frac{3}{8}$ in. and $5\frac{1}{4}$ in. with a 16-in. stroke. Air cylinders 10 in. and 14 in. with a 16-in. stroke. Air pressure, 70-75 lbs. per square inch.

At the lower levels in most mines the water is dealt with by means of electric or compressed air pumps. At the Robinson Main shaft all the mine water from the sixth level is pumped up to the surface by means of one Riedler electric pump. At the City and Suburban the same work is done by means of a 3-throw electric pump.

CHAPTER VII

SURFACE EQUIPMENT OF THE MINES

A WELL-DESIGNED arrangement of surface works and plant is of the first importance for the most economical treatment of ore after it has been mined. The best equipment is, of course, that which will at the least possible cost provide for the handling and transportation of the ore and of the mill products. Naturally the position of the various works and plant, such as rock-breakers and sorting-floors, tramways, mill, cyanide works, etc., depends in some degree on the configuration of the surface as well as on the extent of the property. Mining areas on the Rand, however, neither present great natural advantages nor oppose any considerable difficulties to the laying out of surface works. Thus it is generally possible to find a mill site which, on the one hand, is not too far removed from the main hauling shafts, and on the other is sufficiently elevated to allow of the tailings being led off by gravitation to the settling pits. With regard to tailings, however, it is, as a rule, expedient to lift the pulp coming from the mill by means of pumps or bucket-wheels before they are settled for treatment. In some cases, also, the ore is trammed considerable distances to the mills, and mechanical hoists or elevators are often necessary. With regard to the water required for mills, the heavy rainfall during the summer season affords an ample supply, it being only necessary to conserve the water draining from the hill slopes in large dams, whence it is pumped up to suitably situated reservoirs.

PLATE V.—HEADGEAR (VERTICAL SHAFT) AND SORTING-HOUSE, NEW PRIMROSE MINE.

In the present chapter we propose to give a short account of the nature of the surface plant in use on the Rand, describing the various mechanical arrangements for hoisting, screening, breaking, sorting, and transporting the ore. We shall also have occasion to refer to winding engines, pumps, air compressors, electric installations, etc. The details of the metallurgical plant and processes are reserved for another chapter.

Mining Plant.

The plant which is directly associated with mining operations proper, and is necessarily located in the immediate vicinity of the main working shafts, comprises headgears, winding and pumping engines, air compressors, and small electric installations. There are generally independent steaming plants at each working shaft, supplying all the engines at that shaft, but in some cases, where the mill is close to the main shaft, one set of boilers is made to serve the mill, shaft engines, and a variety of other plant.

Headgears.—Among the headgears in use there is naturally great diversity of design, arising both from the nature of the shaft to be equipped and from the scope afforded to the preferences and opinion of individual engineers. Some of those erected within the last few years are very elaborate and costly, not only serving the primary purpose of hauling and hoisting plant, but also making provision for ore-breaking and sorting, and embodying storage bins of large capacity, which of course form no part of the essential structure of headgears. Five or six years ago simple steel frames from 20 to 35 feet in height were the rule. These were very well adapted for vertical shafts and hoisting with cages, when the ore was trammed directly to the mill bins; and, at a time when the nearest railway terminus was 300 miles from Johannesburg, they were less expensive to import than timber. Now however, they are seldom if ever used, since timber of large scantling can be delivered at reasonable rates, while, in any case, the advantages of timber in such structures would, in the present position of the industry, far outweigh any possible difference in first cost.

One of the largest and most completely equipped shaft-head structures is that at the City and Suburban Main Incline, which covers an area of about 115 feet by 20 feet, and is 60 feet in height. The principal timbers are 14 inches square, the main bracing being 12-inch square pine. The ore skips, which weigh 30 cwt. and have a capacity of 3 to 4 tons each, dump automatically on to the grizzleys[1] near the top of the headgear. The fines run down a shoot under the grizzleys into ore bins of 100 or more tons capacity. The coarse rock from the bottom of the grizzleys is conducted by a shoot to a circular sorting-table, where the waste rock is picked out by hand and thrown into a waste bin to be trammed out and dumped. The pay rock left on the sorting-table is automatically conveyed to a Gates ore crusher, erected immediately under the sorting-floor, and discharging into the main bins into which the fines have already passed. On the floor below the sorting-floor is the gear for working the ore crusher and the sorting-table. In the front part of the head-gear is a second Gates crusher, that can be used as an alternate in case of the other being under repair. Each breaker has a capacity of 45 tons per hour. The shaft having three hauling compartments, there are, of course, three skip-ways in the headgear; the third skip of 30 cwt. capacity is used only for hauling rock broken in dead work, such as sinking and cross-cutting, and dumps directly into a waste bin from which the rock is trammed to the waste heap. The development skip is hauled by an independent engine.

At the Crœsus Main Incline the skips dump on to grizzleys over a shoot leading to a fines' ore-bin, as is usual, but the coarse ore is deposited on a flat sheet at the foot of

1 "Grizzley," an inclined iron or steel bar grating for screening ore.

PLATE VI.—HEADGEAR (VERTICAL SHAFT) AND ENGINE-HOUSE, FERREIRA MINE.

the grizzleys, where it is washed and then shovelled over the edge of the flat sheet on to a travelling belt for sorting. This belt, which is some 50 feet in length, passes between two wash-hoppers, into which the barren rock, picked out, is thrown. The further end of the belt is directly over a Gates rock-breaker, which receives the ore as it drops from the bend of the belt. Under the rock-breaker is a second ore bin holding the broken ore. This bin is of about 100 tons capacity. The fines' bin at the other end of the belt under the grizzleys has a capacity of about 180 tons. The dead rock coming from the mine is dumped into a separate waste-bin, from which it is trammed along a gangway to the waste heaps. The same gangway is also used for the sorted waste.

Among other shaft-head plants that embrace screening and sorting and ore-breaking appliances, are those on the George Goch and Meyer and Charlton. Many headgears are constructed with simply waste- and ore-bins and grizzleys. At the Ferreira the screening and sorting are done on the headgear, but the ore-breaking takes place in the battery.

Waste-bins in headgears are usually small, and are placed high, so that they discharge into trucks on an elevated tramway 15 or 20 feet from the ground, leading out to the waste dumps. The ore bins vary from 50 to 300 tons total capacity.

Self-dumping skips have been adopted universally. The device underlying nearly all arrangements for attaining this object is illustrated in Fig. 62. The back wheels of the skips are made two or three times as broad as the front wheels, or a second wheel is keyed on outside the ordinary wheel. Near the top of the headgear, where the skip is to be dumped, is fixed a pair of rails on to which the back wheels run, while the front wheels, being narrow, continue on the inner or shaft rails. Looked at in side elevation it will be seen that the two pairs of rails diverge, so that the respective pairs of wheels take a different course, the back wheels running to a higher level than those in front of the skip. As a result the skip tips in such a way as to discharge its contents. In some cases the outer rails are set at a higher inclination than the

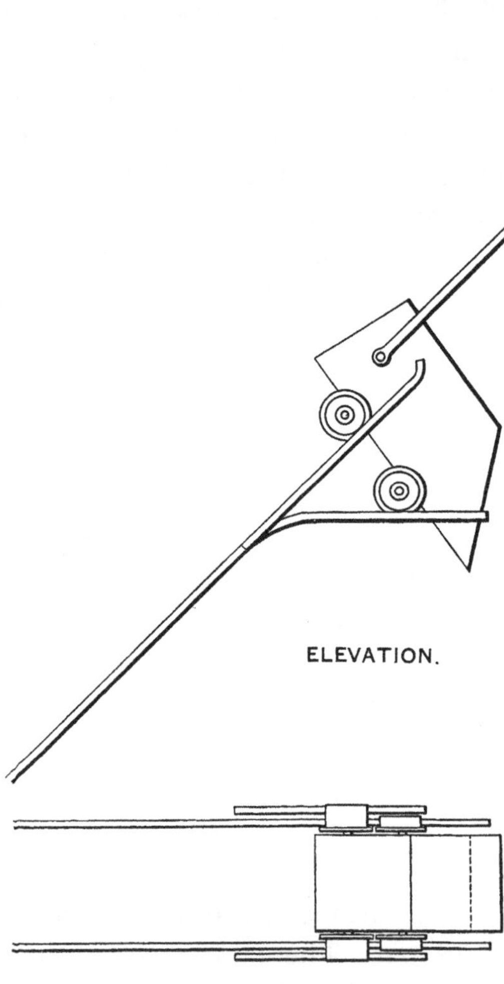

FIG. 62—Plan and Section of a Self-dumping Skip.

shaft rails, in others the shaft rails are bent downwards as in the figure, the result being, of course, the same in each case.

Winding Engines.—The winding engines generally in use are of the double cylinder type, geared to hoist at moderate speeds in the shafts, but for many of the deeper shafts direct acting engines are used. At the Langlaagte Royal a compound

PLATE VII.—HEADGEAR (INCLINED SHAFT), ROBINSON MINE.

winding engine has been erected at the new main shaft. It has two high pressure and two low pressure cylinders, 12-inch and 24-inch diameter respectively, with a 36-inch stroke. The drums on all winding engines have independent clutch and break gear for convenience in hauling from any level in either compartment of the shaft. The use of indicators in the engine-room for marking the position of the skips in the shaft at any moment is rendered compulsory by the Government Mining Regulations.

One of the largest winding engines on the fields at present is that at the City and Suburban main inclined shaft. It is of 180 horse-power, and is direct-acting, with cylinders of 16 inches diameter, and a stroke of 5 feet. It has two independent drums 9 feet in diameter, which at 100 revolutions wind at a speed of nearly 3000 feet per minute. It is fitted with post-breaks to the drums and band-breaks on the crank-discs. The engine used for hauling in the third or development compartment of the same shaft is geared, double cylinder, of about 80 horse-power. Direct-acting engines will be erected at most of the deep level shafts. Those at the Crown Deep Nos. 1 and 2 shafts have each a pair of 20 by 48-inch cylinders, with Corliss valve gear. Flat wire ropes are used, the leverage of the winding reel increasing from about 2 feet with the skip at the bottom of the shaft to a radius of about 5 feet as the rope is wound. There is both a post- and a band-break to each reel.

At the Rose Deep one of the winding engines is a double cylinder of 12-inch diameter, and with a 28-inch stroke, geared; the other is direct acting, with 18 by 36-inch cylinders.

Pumping engines for Cornish pumps are in most cases compound and condensing, and are generally arranged tandem. Single-cylinder engines are also used. Pumping engines are always geared down, and have a crank allowing of a variable stroke in the pumps. Both Corliss and ordinary slide-valve gears are used, with governors and automatic cut-off gear.

Air Compressors.—It has already been stated that very few of the mines in the district are without an air-compressing plant, a fact which is due in great measure to most of the development now carried on being below the limits of the soft "red" rock. The capacity of compressors in use varies from 4 to 30 drill power,[1] and the patterns are of every description, from the simple straight-line compressor to a variety of compound, duplex, vertical, and horizontal types. Among the most notable plants now working are those at the City and Suburban, the Simmer and Jack, and the Van Ryn.

The City and Suburban compressor is of the Riedler type, with compound steam and air cylinders, arranged vertically, the high and low pressure air being over the high and low pressure steam respectively. The plant includes intermediate cooler, surface condenser, and two receivers. The actual capacity is about 21 drills. The steam and air cylinders are the same, 19 inches and 30 inches, the stroke being 36 inches. The chief peculiarity of the Riedler compressors is that the closing of the air valves is not automatic, but is controlled by a positive motion derived from eccentrics or other means. The advantages claimed are—

1. Simplicity, since by allowing of larger valve openings the number of air valves can be reduced, and
2. Large capacity for given size, since the precision and absence of shock in the action of the valves allows of greater piston speed.

The City and Suburban plant, owing to some defects in the details of construction, has not worked up to expectations.

The compressor ordered for the Robinson is also a Riedler,

[1] It should be noted that owing to the high altitude of the Witwatersrand, 6000 feet above sea-level, compressors perform only about 80 per cent of their normal duty at sea-level. Allowance is always made for this in specifying compressor plants.

Fig. 63.—King-Riedler Air Compressor in use at the *Robinson*, *Crown Deep*, and *Rose Deep*. Steam pressure, 120 lbs.; air pressure, 80 lbs.; revolutions per minute, up to 90; capacity, 30 drills (No. 13 Slugger). King's patent engines fitted with Corliss valves and steam-jacketed. Riedler's patent air cylinders and valves; water-jacketed. Cut-off operated by combined speed and pressure governor.

two-stage, of 25 drills' capacity, but the engine is a King Compound. This type of compressor is being put up at several of the deep level mines.

FIG. 64.—The "BURLY" COMPRESSOR in use at the *Nourse Deep*. Indicated horse-power, 350; revolutions per minute, 70; steam pressure, 120 lbs. Compound steam cylinders with Corliss valve gear. Compound air cylinders with inter cooler steam cylinders both jacketed with live steam.

The Simmer and Jack compressor at No. 1 Deep Level shaft is a plant by Walker and Son. It has compound steam and two-stage air cylinders of the following dimensions:—Steam—17 inches and 34 inches by 48 inches; Air—19 inches and 30 inches by 48 inches. The arrangement is horizontal. A surface cooler and condenser are attached. This compressor has not yet been worked up to its full capacity of 25 drills; it has, however, worked well as far as it has been tried.

The Van Ryn plant, ordered for 20 drills, was supplied by the Ingersoll-Sergeant Drill Company. It is compound and condensing, with Corliss steam-valve gear. Steam cylinders 20 and 36 inches, air cylinders $19\frac{1}{4}$ and 31 inches, stroke 42 inches. This plant runs regularly and well, its capacity being considerably above that specified. The arrangement is horizontal.

Among other types of compressor represented on the Rand are those of the Rand Drill Company, Yates and Thom, and the Hirnant.

Compressed air is used chiefly for supplying power to rock-drills, but in many of the mines pumps and small pit-brow engines are also worked by compressed air.

Electric Plant.—On most mines the shaft plant includes a dynamo for electric lighting of underground loading stations and cross-cuts, and of headgears, sorting-floors, tramways, etc., on the surface. In some cases also there are generators for electric transmission of power to underground pumps and winding-engines. Dynamos in shaft installations are generally run by a simple high-speed engine.

Boilers.—On account of the immense power utilised on the fields and the large consumption of fuel entailed, the efficiency and economy of steam plant are naturally questions to which great attention is directed. At the present day on the Rand all the ordinary means of retaining and utilising the heat produced are resorted to, such as feed-water heaters, economisers, the proper setting and building in of boilers, return flues and tubes, etc. Owing to the comparative inferiority of local coals and the high percentage of ash they contain, it has been found advisable to allow a somewhat larger grate area than is usual elsewhere, while combustion is promoted by high smoke-stacks sometimes 150 feet in height. The smoke-stacks for large boiler plants are constructed of steel plate, and rest on a masonry or brick

FIG. 65.—HORIZONTAL MULTI-TUBULAR UNDER-FIRED BOILER. A type largely used in the Rand. Diameter, 66 inches; length, 16 feet; heating surface, 1500 square feet; grate surface, 33 square feet; total weight, 15 tons; nominal horse-power, 125; evaporation of water, 4500 lbs. per hour.

foundation. Four or eight guy-ropes are used for staying the stack.

In the early stages of the industry, when ideas of economy were often crude, boilers of the locomotive type, probably more wasteful than any, were almost universally employed. These have, however, been gradually replaced by more efficient types, such as the Babcock and Wilcox water-tube boiler, and externally fired multi-tubular return boilers. The latter was the first class of economical boiler introduced on the fields, and probably boilers of this type, supplied by a number of different makers, are still used to a greater extent than any other. They afford large heating surface, since the furnace gases, after passing under the boiler, return through internal tubes and are again conducted to the smoke-stack end through external side flues. They have also the advantage over internally fired boilers of large grate area. The Easton and Anderson boiler recently introduced on the Rand is of a type which to some extent combines the advantages of both internally and externally fired boilers. In order to counteract the deficiency of grate area these boilers are fitted with mechanical stokers, so that regularity in the addition of fuel is ensured. The type has not yet had any prolonged trial on the Rand. Another boiler lately introduced is the Heine, a water-tube boiler of German manufacture. The Heine is similar to the Babcock and Wilcox boiler, but claims two or three improvements. The end connections of the circulating tubes are formed by one chamber of boiler-plate riveted to the shell, the tubes being expanded into the plate. There is also an internal drum inside the shell into which the feed water passes before it joins the mass of the water in the boiler. The bulk of the insoluble lime and magnesia salts formed is precipitated in this drum, and can be readily blown off before it settles in more inaccessible parts of the boiler. These boilers are now in use at the Simmer and Jack and Crœsus Mines, and will also be used at the Nigel Deep and the Central Nigel.

In efficient steam plants the consumption of coal varies from 5 to 7 lbs. per indicated horse power per hour, and the water evaporated is usually from 4 to 5 lbs. per lb. of coal.

Plant for Screening, Ore-Sorting, and Preliminary Breaking.

Grizzleys.—The spaces between the bars of grizzleys vary from one inch to two inches, the narrower spaces being adopted where very close sorting is to be effected, or where the rock-breakers are set to break small. The inclination given to grizzleys is from 35° to 45°. From 38° to 40° is about the best angle.

Rock-breakers.—The rock-breakers in use are, with few exceptions, either of the Blake-Marsden or the Gates type, the latter being now most in favour. Although many of the mills still contain rock-breakers, it is customary now to erect them in a separate house, either at the headgear or in some position intermediate between the mine and the mill, and to have an ore-sorting plant in the same building. In this way the rock-breakers can be kept nearer the ground, and are less expensive to erect, while the mill structure is relieved of some very heavy machinery in an elevated position. Further advantages are that much greater capacity can be obtained in the battery ore-bins, and that the broken rock and fines are better distributed and intermixed.

Rock-breakers are usually set to crush to a maximum size of about 2-inch cubes. In some cases, however, they are set closer, or two are used in series—one breaking coarse, the other small. There seems to be little doubt that when the rock is broken small enough to pass a 1-inch or 1½-inch ring, the duty of the stamps is considerably increased, although we have not been able to obtain actual figures showing the degree of advantage obtained. Mr. Osterloh, the manager of the May Consolidated, who has been experimenting in this direction, informs us that he has obtained very satisfactory results. The wear of the head of the Gates breakers used, which would otherwise result in a gradual increase in the size to which the rock was broken, is compensated by a replacable cast steel ring fixed on the wearing part of the head.

Sorting.—Ore-sorting, which on the Rand consists in picking out quartzite from the ore before it is crushed, is carried out as already mentioned, either at the headgears or in an intermediate rock-house. In exceptional cases, but only as a temporary expedient, sorting is done on the rock-breaker floors of the batteries. The appliances in use for spreading out the ore, for convenient and efficient picking, may be classed as follows :—

1. Inclined travelling belts.
2. Inclined tables, with a longitudinal reciprocating motion.
3. Circular rotating tables.
4. Simple floors.

Travelling belts are used at the New Primrose, Crœsus, and the Simmer and Jack. They generally consist of consecutive hinged plates or trays with side flanges, the motion being imparted by means of a hexagonal or octagonal wheel round which the belt is taken. The faces of the wheel fit the separate plates of the belt.

Fig. 66 is a diagrammatic representation of the sorting house at the Roodepoort United Main Reef, where a reciprocating table has been introduced by Mr. Goodwin the manager. Here the rock-house is some distance from the shafts, from which the ore is trammed along an embankment. A A represents the level of the embankment, the tramway extending into the rock-house as indicated. The trucks are turned over by means of the wheel tipper B on to a stationary grizzley G, above the fines' ore bin. G2 is also a grizzley, but it is attached to the table T T, and both are hung so that they can be given a longitudinal movement. This movement is imparted by means of a ratchet wheel or cam C rotating against a tappet on the underside of the table. The table and lower grizzley are thus thrust in the direction indicated by the arrow, until the tappet being released they fall back against a block D with a jerk. By this means the rock is made to travel down the table where a couple of boys on each side pick out the lumps of quartzite before the ore passes into the rock-breaker R. The rock-breaker discharges into the ore-bins. The table is 2 or 3 feet wide and 14 feet long, and about 20 per cent of waste rock is

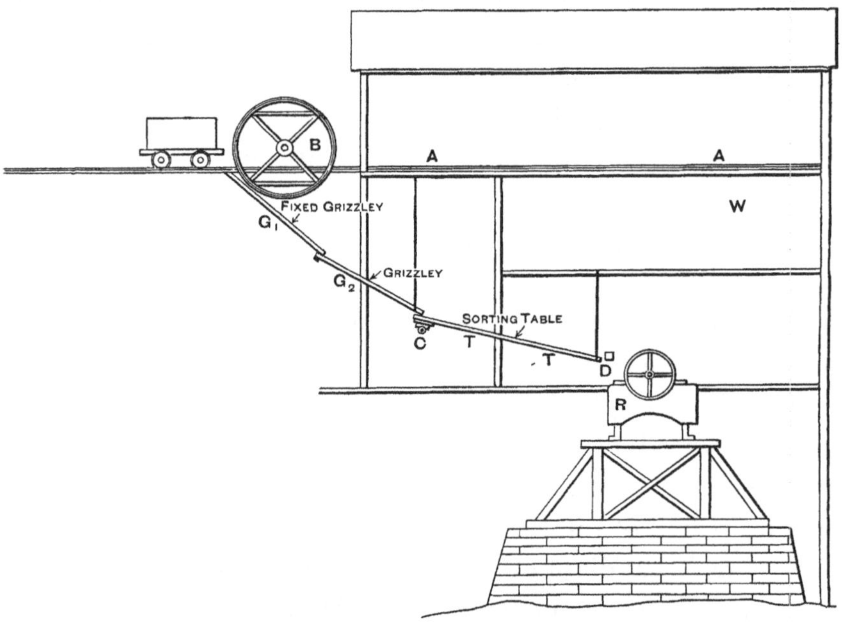

FIG. 66.—Diagrammatic Section of the Sorting-House at the Roodepoort United Main Reef Company's Mine.

picked out. W is a tank with a pipe and hose leading water to the table for washing the rock to be sorted.

Circular tables are in use at the Wemmer, City and Suburban, Metropolitan, Langlaagte Royal, etc. The City and Suburban table, which is represented diagrammatically in Fig. 67, has a diameter of 25 feet. The ore from the grizzleys is delivered through the shoot S on to the outer rim of the table, 2 feet wide, on which the ore is spread for sorting. A slow motion, given to the table in the direction indicated, then brings the ore past the sorting hands, who pick out the quartzite, and throw it into the waste hoppers in front of them. After the rock has passed the sorters, the ore on reaching the rake R is forced over the edge of the table into the mouth of the Gates crusher G, which delivers it to the ore-bins in the lower part of the headgear. The vertical shaft or spindle of the table is rotated by means of a worm

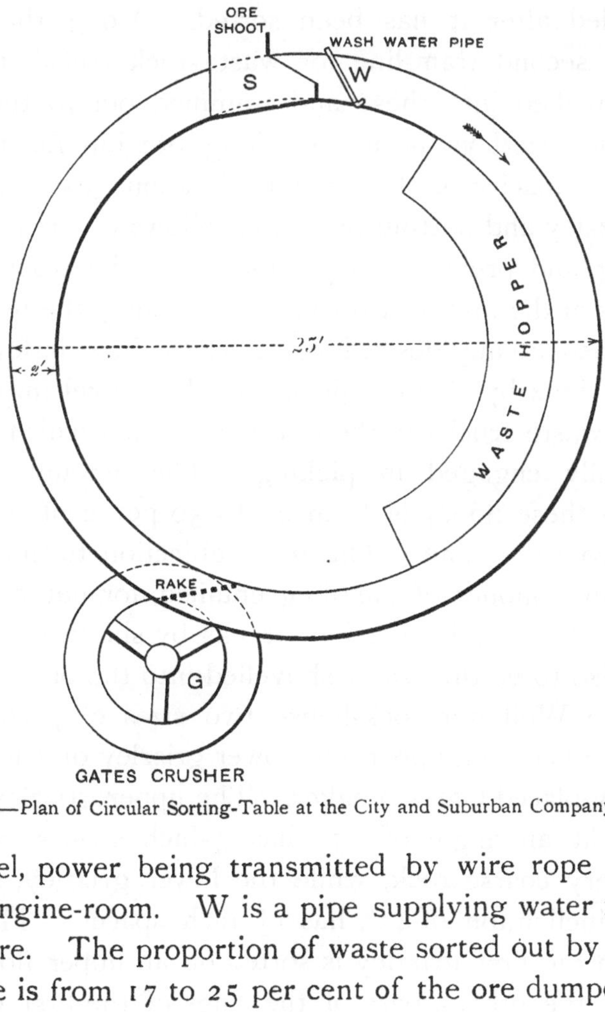

FIG. 67.—Plan of Circular Sorting-Table at the City and Suburban Company's Mine.

and wheel, power being transmitted by wire rope from the battery engine-room. W is a pipe supplying water for washing the ore. The proportion of waste sorted out by means of this table is from 17 to 25 per cent of the ore dumped in the grizzleys at the top of the headgear.

The sorting arrangements at the Ferreira are as follows:—The skips hauled to the top of the headgear dump automatically into a pair of shoots under which trucks are run on a tram-line extending along an upper floor, 60 feet in length. Along the other side of this tram-line are grizzleys, the tops of which are flush with the upper floor, while at their lower end is the sorting-floor, also 60 feet in length by 10 feet in width. On the other side of the sorting-floor, and extending along its whole length, is a series of ore-bins into which the coarse ore is shovelled after it has been sorted. Along the sorting-floor is a second tram-line for waste-rock trucks, the waste being shovelled into these and trammed out to the dumps. Under the grizzleys is another long ore-bin for the fines. The total capacity of the ore-bins is about 300 tons. The large grizzley and sorting-floor area allows of great freedom in picking, and prevents congestion, since the trucks receiving ore from the mine-skips can be run along the upper floor and dumped at any desired point on to the grizzleys. The sorting is done by about eight boys, who shovel into the ore-bins and waste trucks in the intervals during which they are not actually engaged in picking. The amount of waste sorted on these floors is from 40 to 50 per cent of the ore coming from the mine. The great objection to the Ferreira floors is the amount of handling entailed, for not only is the waste handled and loaded into trucks by shovelling, but the ore has also to be thrown or shovelled into the ore-bins.

At the Wolhuter rock-house two sizes of grizzleys are used, there being an upper and lower grizzley on each side of a centrally-placed Gates breaker. The upper grizzley of each pair, set at an angle of 35°, has 3-inch spaces, and only retains very coarse rock, while the lower grizzley, set at an opposite inclination of 42°, has $1\frac{1}{2}$-inch spaces. The coarse rock from the first grizzley is sorted on an upper floor, while that from the others, part of the fines of the first grizzleys, falls on to a lower floor, the fines from the second grizzleys passing direct to the ore-bins. The mouth of the rock-breaker is on a level with the lower sorting-floor, and so placed that the sorted ore from all four grizzleys can be easily shovelled into it. The waste can also be shovelled into conveniently placed hoppers. An average of about 33 per cent is sorted out at the Wolhuter. As pointed out by Mr. Britten, the manager, who designed this plant, there is an advantage in keeping the very coarse rock separate from the smaller lumps in sorting, since a big lump of rock frequently falls on the top of the smaller ore, and prevents effective washing as well as expeditious picking.

No clearer exposition of the advantage of ore-sorting could be given than that contained in table H, which is extracted from a published report of Mr. J. H. Johns of the Ferreira, to whom is due the credit of the first systematic effort in this direction. The degree to which sorting is possible depends on the proportion of quartzite accompanying the ore, and the advantage is of course greater with small reefs, the mining of which entails the breaking of much waste rock in the stopes.

The proportion of waste rock handled being extremely variable, it is impossible to institute a comparison of the efficiency of the plants in use on different mines. While at the Wemmer and Ferreira 40 per cent to 50 per cent is sorted out, it would be impossible, for example, at the Simmer and Jack, where the reefs are thicker, to eliminate more than about 25 per cent.

TRANSPORT OF ORE.

The transport of ore on the surface from shaft to battery is always effected on tramways, the gauge of which is as a rule 18 inches, the ore-trucks holding from 16 to 20 cubic feet of broken ore.[1] The rails generally weigh from 14 to 16 pounds per yard.

[1] The number of cubic feet estimated per ton of broken ore varies somewhat, according to the opinion of individual operators, which might to some extent be expected since the bulk of a ton differs according to the size to which the ore happens to be broken. The usual allowance is from 20 to 22 cubic feet.

TABLE H.

Ferreira Gold Mining Company, Limited.

Statement showing advantage of sorting.

Period	Ore mined.	Waste Rock sorted.	Ore milled.	Percentage of Waste Rock picked out.	Value of Ore mined.	Value of Ore milled.	Including value of Ore milled due to sorting.	Value of additional Gold won due to sorting.			Direct saving due to sorting.		
	Tons.	Tons.	Tons.	Per cent.	Dwts.	Dwts.	Dwts.	£	s.	d.	£	s.	d.
Year ending March 1892 . .	57,724.5	9,017.5	48,777	15	19.38	22.69	3.31	28,899	17	4.8	6,825	4	1.84
Half-year ending September 1893 .	35,253	11,217	24,736	31.19	19.7	27.95	8.25	36,526	14	9.6	9,360	3	9.65
,, ,, March 1894 . .	34,259	11,782	22,477	34.36	22.491	33.495	11.004	52,559	17	9.36	10,282	0	9.62
Nine months ending December 1894	66,543	29,753	36,784	44.72	19.33	33.762	14.432	109,547	4	5.392	14,656	13	4
	193,779.5	61,769.5	132,774	...	...	...	...	227,533	14	5.152	41,124	2	1.11

The average assay value of the waste rock picked out in sorting is less than that of the discarded residues from the Cyanide Works.

It will be observed from the above statement that gold of the value of £227,533 : 14 : 5.152 has been won by the company since the inauguration of the system of sorting which could not otherwise have been obtained with present plant in the same space of time.

The company has also been saved the expenditure, after deducting the cost of sorting, of £41,124 : 2 : 1.11 in not having to transport, mill, and cyanide 61,769.5 tons of waste rock.

PLATE VIII.—Transport by Endless Rope Haulage at the New Primrose Mine.

If the mill is in close proximity to the shafts native labour is often employed in tramming, but for longer distances it is usual to have mules or mechanical haulage. Sometimes, depending on the grade, power need only be applied in returning empty trucks. In one instance, viz., at the Crown Reef, electric locomotives are used, while at the Simmer and Jack a switchback tram is in use between one of the shafts and the mill. Endless rope haulages have only of late years been very generally adopted, though one was in use at the Durban Roodepoort adit-level some four or five years ago. Mechanical haulages have now been applied at the Primrose,

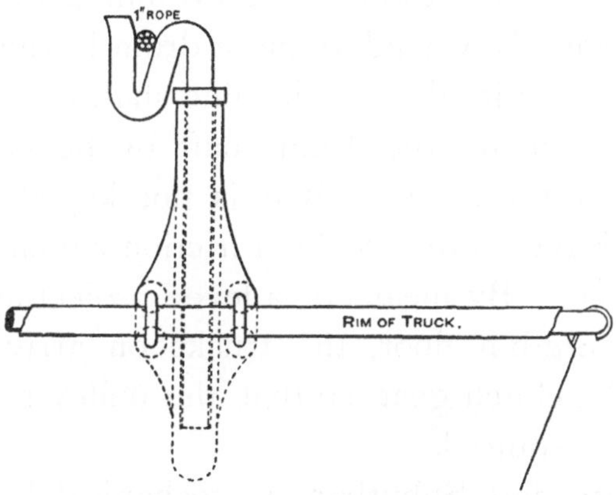

FIG. 68.—Jockey used in Endless Rope-haulage.

Jumpers, Geldenhuis Estate, Wolhuter, Orion, and many other mines. The endless rope-haulage from the Jumpers Mine to the mill is 2700 feet in length. Sometimes a trailing rope is used, but more often the rope is suspended, the trucks being fastened by a V-shaped clutch or "jockey," turning in a socket which grips the rope as sketched in the accompanying figure 68.

The Crown Reef tram-line is laid with 20-lb. steel rails at a 24-inch gauge. The trucks are side-tipping, and of 30 cubic feet capacity. The total length of tram-line is nearly one mile. The two locomotives, each carrying a 20 horse-power motor, are operated on the overhead trolley system, power being transmitted by dynamos with a voltage of 550.

Since it is only in exceptional cases, as at the Geldenhuis Estate, that sufficient difference in ground-level exists between mine and mill site to allow of trucks being run on a level embankment to the top floor of the mill, some mechanical method of raising the ore has in most cases to be resorted to. The means generally in use are inclined planes, with hauling gear worked from the battery shafting, and vertical elevators actuated from the same source. Thus, at the Simmer and Jack the ore trucks are hauled up an incline on trestles from the end of the surface tram to the ore-bin floor of the mill. The haulage rope is wound round a drum in the battery, the power being transmitted from the line-shafting to a counter, which in turn actuates the drum shaft by bevel gear. The driven pulley on the counter shaft is not keyed, so that this shaft can be thrown into gear by a friction clutch when trucks are to be hauled. By means of a lever projecting above the track on the ore-bin floor, the trucks on arrival automatically detach the clutch gear, so that the pulley runs idle until power is again required.

At the City and Suburban a mechanical haulage, on a tail-rope system, takes the ore-trucks from the headgear-bins round a curve and up an inclined plane to the battery-bin floor. Here the haulage drums are worked through bevel gear by means of two fast pulleys, one with a crossed belt and reverse motion, there being also an intermediate loose pulley for stopping the haulage.

Elevators, into which the trucks are run at ground level, and raised vertically to the tipping-floors, are used at the Durban Roodepoort, Langlaagte Estate, Heriot, and several other mines.

Workshops.

Nearly every mine of any pretensions includes in its equipment engineering workshops, with a more or less complete outfit of machine tools. By this means a great deal of delay, trouble, and expense is avoided, both in the matter of repairs and alterations, and in the construction and erection of new plant, while the further advantage is secured that all work of this kind is done under the direct supervision of the superintending engineer and manager, or their official representatives.

Engineering shops are generally furnished with ordinary lathes, screw-cutting lathes, shaping, drilling, slotting, planing, shearing machines, etc. Carpenters' shops are equipped on a similar scale. The Robinson and Crown Reef workshops are examples of a very complete outfit in most respects, the former even including a small foundry plant. Smiths' shops, which are of course indispensable, contain sometimes as many as ten or a dozen forges, thousands of drills passing through the smiths' hands every day. A common adjunct of smiths' shops is a small steam hammer. Among labour-saving appliances adopted on the Rand is a drill-sharpening machine, by means of which one smith and a couple of boys can finish off from 600 to 1000 bits in a shift.

WATER-SUPPLY.

As explained at the outset of this chapter, the water necessary for milling and cyaniding operations is obtained by conserving in reservoirs or dams the summer rainfall. These dams are made across the vleis or boggy water-courses which occur on or in the neighbourhood of most properties. The natural slope on either side of a vlei is usually very gradual, so that the reservoirs are, comparatively speaking, shallow and of considerable area. The loss from evaporation is probably great, but there is little soakage, as the soil and the rocks below are fairly impervious, while in some cases a natural clayey bed is found under the surface soil. There are very few permanent streams in the district, and such as exist are too small to be of any practical service without storage.

At the Roodepoort United Main Reef, Mr. Goodwin proposes to obtain a battery supply by sinking two shafts and driving to connect. The wet nature of the ground in the neighbourhood of the mill renders the success of this undertaking highly probable. The saving in working costs will be about a shilling per ton, since, under present conditions, the company pays the Princess Company for water at the rate of about 1s. 2d. per ton treated.

Dams are mainly constructed of earthwork, the material being obtained from the bed or sides of the vlei. The earth, generally containing a fair proportion of clayey matter, is carted or trammed to the dam-wall, and tramped in by the animals employed in the work of construction. It is customary to commence by carrying an excavation across the bed of the vlei down to the bed-rock. Most of the dam-walls are built with a core of puddled clay along the central portion, the core being two to three feet in thickness, and extending some feet into the bed-rock. Ample bye-washes have to be provided at the ends of the dam wall, disregard of this precaution having in several instances led to serious damage during the rainy season. The inner face of a dam is usually ballasted at the toe and pitched with stone. The slopes vary from $1\frac{1}{2} : 1$ to $3 : 1$.

The only masonry dam in the district is that now being built by the Rand Mines, Limited, for the Geldenhuis Deep and other properties. This dam will be 40 feet in height and about 1500 feet in length, the capacity being estimated at 700,000,000 gallons.

The capacity of dams on the Rand seldom exceeds 50 million gallons, that of the majority being between 10 and 30 millions.

The distance from which water has to be pumped from the dams to the mill reservoirs, although in some cases considerable, does not as a rule exceed a few hundred yards. The pumps most in use for this purpose are those of the horizontal duplex and 3-throw types, which may be worked by wire-rope transmission, by belt transmission from a small engine in the pump-house, or by electricity. Electricity as a means of power transmission for this and other purposes is daily gaining favour. The water - mains, which are from 6 inches to 12 inches in diameter, according to requirements, lead to reservoirs, generally of a few million gallons' capacity, situated above the mill-site, so that sufficient head is obtained. The reservoirs are partly excavated and partly surrounded by an earthbank, advantage being generally taken of a natural slope, so that in levelling the bed of the reservoir material is obtained for the construction of the walls. It is usual to have a core of puddled clay from bed-rock upwards, at least on the low side of a reservoir, and to pitch the inner slope with stone.

There being no permanent natural supplies, economy has to be observed in the use of water, and in most cases the tailings' water is returned to the dams after settlement of the slimes. A certain loss, which may amount to 20 to 30 per cent, is of course inevitable, but, in estimating the required storage-capacity of a dam, it is by no means necessary to make allowance for the total quantity of water required from day to day throughout the dry season.

Water for boilers is generally taken from the main reservoirs, and is consequently apt to contain fine matter in suspension as well as dissolved impurities. The latter are, however, not present in any serious degree. If mine-water from below the level of oxidation is used it is liable to contain free sulphuric acid, sulphate of iron, and other exceedingly injurious ingredients. The subject of boiler feed-water is now receiving greater attention than formerly. In cases where suspended matter is excessive, filters are used, acid mine water is avoided when possible; purifiers are being adopted, and more careful attention is paid to boilers.

Mill-Motors.

The largest engines used on the fields are, of course, those erected for driving milling plant. Since each stamp requires from $1\frac{1}{2}$ to $2\frac{1}{2}$ horse-power, and since in most cases additions have been contemplated in ordering the motor, engines of several hundred horse-power are not uncommon. Stamps, again, are by no means the only load put on mill-engines, for besides these and the usual accessories to a battery, such as rock-breakers, concentrators, etc., heavy work in connection with the water-supply and the driving of hauling systems are in many cases added to the ordinary duties of a mill-power plant. Thus at the Langlaagte Estate the battery engine, which is compound and condensing, with 22-inch and 32-inch cylinders, and a 4-foot stroke, working at 100 lbs. pressure, drives, besides 160 stamps, 96 Frue vanners (about $\frac{1}{2}$ h.-p. each), a haulage, and three dynamos aggregating 80 horse-power, which transmit power to three No. 6 Gates crushers, and to the cyanide works, and generate light for all the surface works.

The Crown Reef steam-motor is a single cylinder, Corliss condensing engine, with a cylinder measuring 28 inches by 60 inches. It drives at present 120 950-lb. stamps, the dynamos transmitting power to the dam pumps, the locomotives, and the lighting dynamos. The pumps are 3400 feet from the mill reservoir, and have a capacity of 60,000 gallons per hour to a head of 200 feet, requiring 95 horse-power from the motor. The locomotives carry each a 20 horse-power motor. There is in the engine-room a 250 horse-power auxiliary engine for the electric plant in case of a breakdown in the main engine, which is of 750 horse-power. The steaming plant is of about 800 horse-power.

The City and Suburban mill engine, which is probably capable of developing 800 indicated horse-power, is vertical, triple expansion, and condensing with Corliss gear. The

PLATE IX.—BOILERS AND ENGINE IN COURSE OF ERECTION AT THE WOLHUTER MINE.

FIG. 69.

FIG. 70.

MILL ENGINE in use at the *Henry Nourse*. Compound-condensing—King type. Corliss valve gear. Steam cylinders, 19 in. and 30 in. diameter; stroke, 42 in.; revolutions per minute, 75; indicated horse-power, 330. Automatic cut-off in high-pressure cylinder to 70 per cent of the stroke. Hand cut-off in the low-pressure cylinder. Both cylinders steam-jacketed.

high-pressure cylinder is above the intermediate, the low-pressure cylinder being parallel. The dimensions of the cylinders are: high pressure, 20 inches; intermediate, 31 inches; low pressure, 48 inches; stroke, 48 inches. The working pressure is 140 lbs. to the square inch. The load at present consists of 80 stamps, 2 mechanical haulages, a No. 6 Gates crusher, sorting-table, the water-supply pumps, a 40-feet tailings wheel, pumps at the cyanide works, an electric pump raising water from the 6th level at the main incline shaft, and the surface and mine lighting. 80 more stamps will be added in a few months time. The steaming-plant, which also supplies two large winding engines and a 20-drill air compressor, consists of eleven 100 horse-power multi-tubular return boilers connected to one steam main. There are two smoke-stacks 111 feet in height, 6 feet in diameter at the foundation, and 5 feet at the top.

The Robinson new power plant will be one of the most complete on the fields. The boilers are of the multi-tubular return type by Fraser and Chalmers, 14 in number, and 140 horse-power each, equal to 1960 horse-power. The main driving engine is a tri-compound King, of the vertical marine type, 600 horse-power, also by Fraser and Chalmers; cylinders: H. P. diameter 19 inches; I. P. 30 inches; and two L. P. 30 inches each; stroke, 42 inches. The H. P. and I. P. cylinders work on one crank, the two L. P. cylinders on another. The valve gear is Corliss, operated by a governor of the central weight type, the high-pressure cylinder having a range of 0.7. The other cylinders are fitted with hand-adjustable cut-off gear, and have the same range. The peculiarity of the King engine is a triangular connecting-rod, which has a uniform turning motion equal to two cranks. Power is transmitted directly to the line shafting of the mill, which is continuous with the main shaft of the engine. Power is also transmitted by rope belt to the electric counter-shaft, which is fitted with Seymour friction clutches, and thence to four Mather and

Photographed by Davies Bros.

PLATE X.—INTERIOR OF ENGINE-ROOM, CITY AND SUBURBAN MINE.

Platt generators of 100 horse-power each. (120 volts, 620 ampères, 620 revolutions.) The electric power is to be transmitted (1) to the pumping station $\frac{3}{4}$ mile distant, where there are two sets of 3-throw pumps delivering 50,000 gallons an hour (driven by a 60 horse-power motor); (2) to the Riedler pumps in the mine; and (3) to the electric light

FIG. 71.—RIEDLER STEAM PUMP for underground purposes, as used at De Beers Consolidated Mines. Most suitable type for direct driven underground pumps. Have been worked successfully for some years at the Peter shaft near Prague at a constant vertical lift of 1700 feet. Plungers differential $7\frac{5}{8}$ in. and $10\frac{1}{4}$ in. diameter by 30 in. stroke. Normal speed, 85 revolutions per minute. Frequently run at 120 revolutions per minute against a head of 100 lbs. per square inch. Only 2 valves coupled direct to engine shaft.

service throughout the property. There will be erected in the same room the present Corliss 14 by 28-inch engine as an auxiliary, a Rand duplex compressor with compound steam cylinders, and a King-Riedler two-stage compressor to drive 25 drills. The steam cylinders of this compressor are: H. P. 19 inches; L. P. 30 inches; the air cylinders are 20 inches and 30 inches diameter, stroke 42 inches.

DEEP-LEVEL MACHINERY.[1]

(*By L. J. Seymour.*)

For present calculations, taking the flattening of the lodes into account, we may take 3000 feet vertical depth as the greatest which we need discuss, leaving for future consideration the subject of machinery for deeper working, when greater experience will no doubt make operations at a depth of 5000 to 6000 feet seem as feasible as those at 3000 are at present.

The principal machinery to be considered is that for rock-drilling, tramming, pumping and hoisting,—the latter two of which only require to be specially designed, the former in most respects simply being a continuation on a larger scale of existing apparatus.

Rock-drilling.—Should the rock become more dense as greater depths are reached, which is improbable, it would for economic reasons require a wider application of pneumatic rock-drills. The relative value for drilling purposes of human labour and air drills varies in a direct ratio to the density of the rock to be drilled. Where it is soft enough, as, for instance, the "blue ground" at the Kimberley diamond mines, a hole may be "jumped" in to the depth required for blasting in the time that would be required to get an air drill set up and in operation, while the number of changes of bits would make the whole expenditure of time and tools altogether out of proportion to the saving to be effected by the more rapid boring while the drill was actually in motion. With harder rock, such as the "black shale" at the diamond mines, the difference between the two becomes unimportant, and the question for decision is then the cost of equipment,

[1] In compliance with our request, Mr. L. J. Seymour, M.I.M.E., the well-known mechanical engineer, and formerly chief engineer at De Beers, has made the above interesting contribution to this chapter.

with a compressed air plant, versus a regular supply of manual labour at rates comparing favourably with the cost of working the air plant.

In the Transvaal the sources of the labour-supply are so abundant that it appears probable that a good supply of native drillers, at rates somewhat lower than are now paid, will be available for many years, so that the use of air drills will be generally restricted to sinking and driving operations.

For narrow stopes, the difficulty of setting up and working air drills counterbalances the other advantages, leaving this field to hand labour; but when the width of reef is over 3 feet, the air drill may be expected to oust hand-drilling, even where the rock is only moderately hard. Experience with the different makes of rock-drills in confined workings will probably develop a type smaller and lighter than is now used, but requiring air at a higher pressure. It is reasonably certain that the air pressures carried will soon be 90 to 100 lbs. per square inch instead of 70 to 80, as now employed, and that the cumbersome drill-holding apparatus still frequently seen will give way to a lighter form. The effective blow varies directly as the mass of the reciprocating part, and as the square of its velocity, so that increased striking velocity is of greater importance than the weight of the striking parts. Also in slow moving drills the air valve is moved, as soon as it has moved past the middle point in its stroke, to admit the pressure to the bottom side of the piston; and unless the velocity of the piston is great enough, the valve will have moved sufficiently to admit the return pressure before the blow has been struck on the rock, thus partially neutralising the effect. The consumption of compressed air by the present forms of machine drills is enormous, and waste by leakages increases directly as the pressure increases. Better attention to lubrication and fittings will diminish the latter loss, but the only remedy for the former appears to lie in some form of air cylinder which shall use the air expansively. Indeed, such machines have already been made, and it is probable that the prejudice against their use will be overcome as the type becomes perfected. Large central stations for compressing air, under competent supervision, instead of the numerous small ones now so commonly seen, will decrease the cost of the supply, so that it will be much less expensive to use than at present. Heating the air before use in underground drills is impracticable, the disadvantages in working outweighing the economic gain.

Although electric drills are occasionally made, it is not probable that they will be used in mining to any extent, since they are unsuited to the rough usage accorded to drilling machines—the advantage that wires can be laid more easily than air mains being outweighed by the heat generated while working, and the absence of the ventilation which is afforded by the air drill.

Mechanical tramming is of course the same at all depths, the motors and methods adopted requiring no changes from those in use at lesser depths.

Pumping from a depth of 3000 feet does not appear to present any serious difficulties. Some of the types of pumping machinery suitable for 1500 feet should be superseded by others now known to be efficiently used in other mining regions. While for sinking purposes to moderate depths, the geared Cornish pumping plant with a single wooden spear-rod, having poles attached at intervals of 300 feet, is undoubtedly the simplest and most convenient arrangement, it becomes too costly and cumbersome for depths much greater than 1500 feet.

Assuming that sinking operations are carried on fairly constantly during the entire life of a mine, some simple type of direct acting pump, without exposed working parts, and operated by compressed air, appears to be the most convenient type to use at the lowest levels for raising the water to the main pumps. Such a machine may be hauled up quickly, sufficiently out of harm's way, to allow blasting by battery to take place without injury to the pump and piping, and again quickly lowered and set to work. An exceedingly simple, cheap, and convenient apparatus for this purpose is the "Pohle" air-pumping arrangement. Briefly described, it consists of a pipe whose lower end dips in the water to be pumped, and whose upped end leads into the suction tank of the permanent pump, located 50 to 100 feet farther up the shaft. A small compressed air pipe leads down outside the water pipe, its lower end terminating in a U-bend, having its opening turned up into the centre of the water pipe. When compressed air is turned on, slugs of water and air are alternately carried up without the use of valves or other mechanism. Sand and dirt do not affect its operation, and where a large quantity of water has to be lifted quickly to a height not exceeding 100 feet it is a simple and certain arrangement. One that the writer saw recently was used for constantly lifting 200,000 gallons of water a vertical height of 40 to 65 feet per day. The water pipe was 3 inches internal diameter, the air pipe 1 inch, and the air pressure 55 to 60 lbs. per square inch. The loss of heat by the expansion of the air did not appear to materially affect the temperature of the water raised, showing to some extent that the consumption of air could not have been great.

For pumping from permanent stations, the amount of water to be pumped has an important bearing on the type of pumping machinery best suited for the position. Where the quantities of water are much greater at the upper levels than at the lower ones, as in many of the Johannesburg mines, small direct-acting or geared pumps operated by compressed air or electricity may be employed to raise the water to the main pumping station. At this point a complete plant, always in duplicate, and of a capacity sufficient for all the mine water and any surface water which may find its way into the mine during an exceptionally rainy period, should be placed in a chamber of sufficient size to allow attendance and repairs to be carried on with ease and despatch. If the quantity of water to be pumped is over 50,000 gallons per hour, steam is the best driving power under average conditions, especially if the steam pipes may be carried down an upcast shaft. The steam pipe should be small enough to give a velocity of 100 feet per second when working at nearly maximum power, the heat generated by the friction of steam in the main serving to evaporate any water of condensation, so that the loss by friction is practically counterbalanced by the gain in dryness of steam. A steam receiver of ample size should be placed at the cylinder to give a steady pressure on the piston, and maintain a uniform flow in the main. Obviously the engine should be compound and condensing, both on account of economy and for disposing of exhausted steam. For a depth of 2000 feet the appreciation of temperature of the water pumped would, in a fairly good plant, not exceed 20 degrees Fahrenheit when all the steam is condensed by the mine water.

The cost of erecting the geared Cornish type of pumping machinery is often the most serious item to be borne. Heavy foundations are needed at the surface to withstand the shocks of reciprocation. The moving weight of the rods and poles of four sets of Cornish pump-work each 14 inches diameter, with a 10-feet stroke, working to a depth of 1000 feet, is as much as 70 tons, leaving out of the question the weight of the balance bobs and pitman. Chambers must be excavated for access to the pump valves and for the suction tanks, at intervals of 200 to 300 feet in the shaft. Heavy catchers must be put in every 500 feet to save the apparatus from serious wreckage should an accident occur to the pitman, or the connection to the nose of the bob give way. The moving rod also is a source of danger to those climbing up and down the shaft unless it is bratticed off. The delivery pipes need to be as large in diameter as the plungers themselves, since the

PLATE XI.—CORNISH PUMP.

Diameter of ram, 14 inches; stroke, 10 feet. Greatest working speed, 10 strokes per minute. Valve chambers each with 4 flap-valves, leather-faced.

delivery of water in the mains is only taking place half the time. All these costs are so serious that, despite the advantage of having the driving mechanism at the surface, where it can be easily taken care of, it will often be found that the interest on the money expended would be greater than any economy over pumping machinery operated by electricity or compressed air. Recently it was found, in a well-known American gold mine, to be actually cheaper to buy a first-class compound pumping engine, and install it in complete working order, than to erect in position a complete Cornish pumping plant lying already at the mine. This would be especially pertinent to African mines, where skilled labour is paid at a high rate of wages.

Given a well-designed mill engine of ample power, constantly running, with a reliable auxiliary engine, the question of electricity *versus* steam for driving a main pumping plant situated far underground becomes one for careful consideration, although electric motors in South Africa labour under the disadvantage of frequent and severe thunderstorms, and the only safety from serious breakdowns lies in having spare motors and parts available for quick replacement. Electricians object to handle currents much over 200 volts underground, although in many instances currents of 500 and even 700 volts are now employed. Where the shafts are well timbered and dry no especial difficulty obtains in laying and protecting mains, except where a malicious or mischievous "boy" may attempt to cut or destroy them, in which case, if the higher voltages are used, the same boy does not usually repeat the experiment.

Electric motors suffer greatly from intermittent working shocks, so that pumps having three rams worked from cranks 120 degrees apart are imperative for this motive power. The earlier types have the motor belted to a countershaft on which a small pinion gears into a spur-wheel on the pump crankshaft, but in later forms, now largely used, the motor is geared directly with the crank-shaft of pumps having Professor Riedler's mechanically controlled valves, by means of which the pumps may be run at a plunger velocity of upwards of 500 feet per minute, or about three times the speed of pumps not so constructed.

The writer recently tested an electrically driven Riedler pump, working against a water pressure of 1300 lbs. per square inch, and its operation dispelled all doubt of the possibility of adopting these pumps safely and economically for regular work under a head of 2000 feet, thus avoiding the numerous stations and attendants necessary where short lifts are in use. A notable instance of this kind is the Riedler pump at the Petershaft near Prague, which has raised water 1700 feet at one lift without difficulty for several years past, while instances of lifts to 1000 feet are very common.

In considering the cost of pumping we may take an actual case which has come under the writer's notice.

The cost of the Cornish pumping plant, consisting of a twin compound, steam-jacketed engine of 40 nominal horse-power, geared to a bob driving four plungers, each of 8 inches diameter and 6-foot stroke, and pumping from a depth of 1000 feet vertical, including two boilers (one for reserve), was roughly £7800, erected in working order. The service of three men, viz. engine-man, stoker, and pump-man, per shift, or six men per day, was required. The average number of gallons pumped was about 6000 per hour, and the colonial coal consumed was 4 lbs. per horse-power, equal to $2\frac{3}{4}$ lbs. of English coal in heating value. The total friction of the plant, calculated by the average steam pressure on the piston, as referred to the actual number of pounds of water lifted, multiplied by the actual head overcome, was nearly 28 per cent, and the displacement of the plungers was 97.6 per cent of the theoretical displacement.

Thus we have :—

Wages of 2 engine-drivers per annum	at 20s.	per shift,	£730	0	0
,, 2 stokers	,, 6s.	,,	219	0	0
,, 2 pump-men	,, 25s.	,,	912	10	0
Total wages			£1861	10	0

Fuel per annum at 26s. per day	£474	10	0
Oil, waste, packing, etc.	260	0	0
Repairs, say 2 per cent	156	0	0
Depreciation at 6 per cent	468	0	0
Interest on the capital expended at 5 per cent	390	0	0
Total cost per annum	£3610	0	0

Such a pump is capable of raising twice the amount of water referred to, but, in general, a plant of this type must be capable of exceeding the average quantity of water raised by 100 or 200 per cent if needed, which makes the initial cost and operating expenses very high. In spite of these high costs, the certainty of action, should the lower pumps be drowned, and its advantages for sinking purposes, are sufficient to cause its adoption largely; but when a depth of 1500 to 1800 feet is reached the initial outlay becomes too great for commercial utility, and other types must be substituted.

Now that electricity has come to play such an important part in driving the various subsidiary machinery in and about a mining plant, its extension to pumping machinery is only a step farther on the beaten track. The addition of an extra generator does not usually mean additional cost of attendance to an existing electrical plant, while the pump-man can readily look after the motor, so that two men per shift are required as against three with the other plant referred to, making a saving in attendance of £700 per annum. One pump at 1500 feet depth replaces five isolated plungers at intervals of 300 feet each. The main need only be one-half the size used for the Cornish pump, since the flow of water is continuous, and the cost of a duplicate pump, held in readiness for an emergency, is not a considerable item. For these reasons it appears that electricity is likely to come widely into use for deep level pumping, unless the quantities of water are large enough to require motors and mains of great size, in which case the steam pumping-engine underground is likely to take precedence.

Hoisting Machinery.—At present there are a number of types of winding plants, differing considerably from one another, each in high favour in certain localities, often to the exclusion of all other types.

In certain of the Western States of America, where rapid winding is carried on from different levels, direct-acting engines with two reels, each operated by a steam clutch and using flat ropes, are in universal use. Lowering is done by the brake, and since the diameter of the reel, when the skip is at the bottom, is frequently as small as 3½ feet, hoisting may go on without the counterbalancing effect of the other cage, even with cylinders of a comparatively small size. The flat ropes are made of as many as fourteen ropes, each of ½ inch diameter, laid side by side, and sewn through and through with fourteen strands of charcoal iron or very mild steel wire of about 14 gauge. Flat ropes, however, can only be used in vertical shafts, as they would be worn too rapidly on an incline.

In other localities large cylindrical drums grooved for round ropes, and driven by steam cylinders of great size, are in fashion, sometimes with one drum on which both ropes wind, one winding on as the other winds off. The greatest disadvantage of this type is the inertia of the drum, especially where a tail rope is not used. Fast winding from different levels is impossible unless the cylinders are out of all proportion to the size of the drums, and the great size of the plant with the excessive foundations necessary bar its general

PLATE XII.—WINDING ENGINE, 20″ × 48″, IN USE AT THE JUMPERS DEEP.

Steam pressure at receiver 120 lbs. per square inch. Cylinders, steam-jacketed. Corliss valves. Cut-off operated by throttle lever, from zero to full stroke. Steam reversing engine with oil cataract. Single drum 8 feet diameter by 5 feet 6 inches wide for round rope.

adoption for commercial reasons. Conical grooved drums mitigate the evil to some extent, but are also costly, and have been the cause of serious accidents. They have the advantage of counterbalancing the weight of the rope, but as this advantage results in decreasing the size of the steam cylinders, the engines are nearly as unwieldy as those having cylindrical drums.

In 1866 a system of winding engines was patented by Mr. S. B. Whiting, now the general manager of the famous Calumet and Hecla Copper Mines, which, while allowing for almost unlimited extension, obviates the necessity for large and expensive drums. The improvement consists in the use of two grooved drums, one directly in line with the other, around both of which the rope from one compartment of a double shaft winds, and after making three round turns is led to the other compartment of the shaft. This arrangement has been almost universally adopted for cable tramway systems, but until lately has not been much used for winding purposes. The engines are preferably placed at the end of the shaft instead of the side, one head sheave being directly over the other, especially when winding always from one level as in coal mines, although by using a guide sheave this arrangement of drums may be placed in the ordinary position at the side of the shaft. To allow of winding from different levels, the rope, after passing thrice round the drums, is led back around a sheave on a moving carriage, returning from thence to the head-gear sheave (see Fig. 72). The track on which the carriage containing the take-up sheave works requires to be half as long as the greatest distance between the upper and lowest levels to be hoisted from, since a movement of one foot in this carriage pays out or takes up two feet of slack rope. The track requires a strong anchorage at its extreme end, and if the changes to different levels are frequent, a steam winch must be connected to the carriage for varying its position quickly. The drums are usually lined with wood blocks to

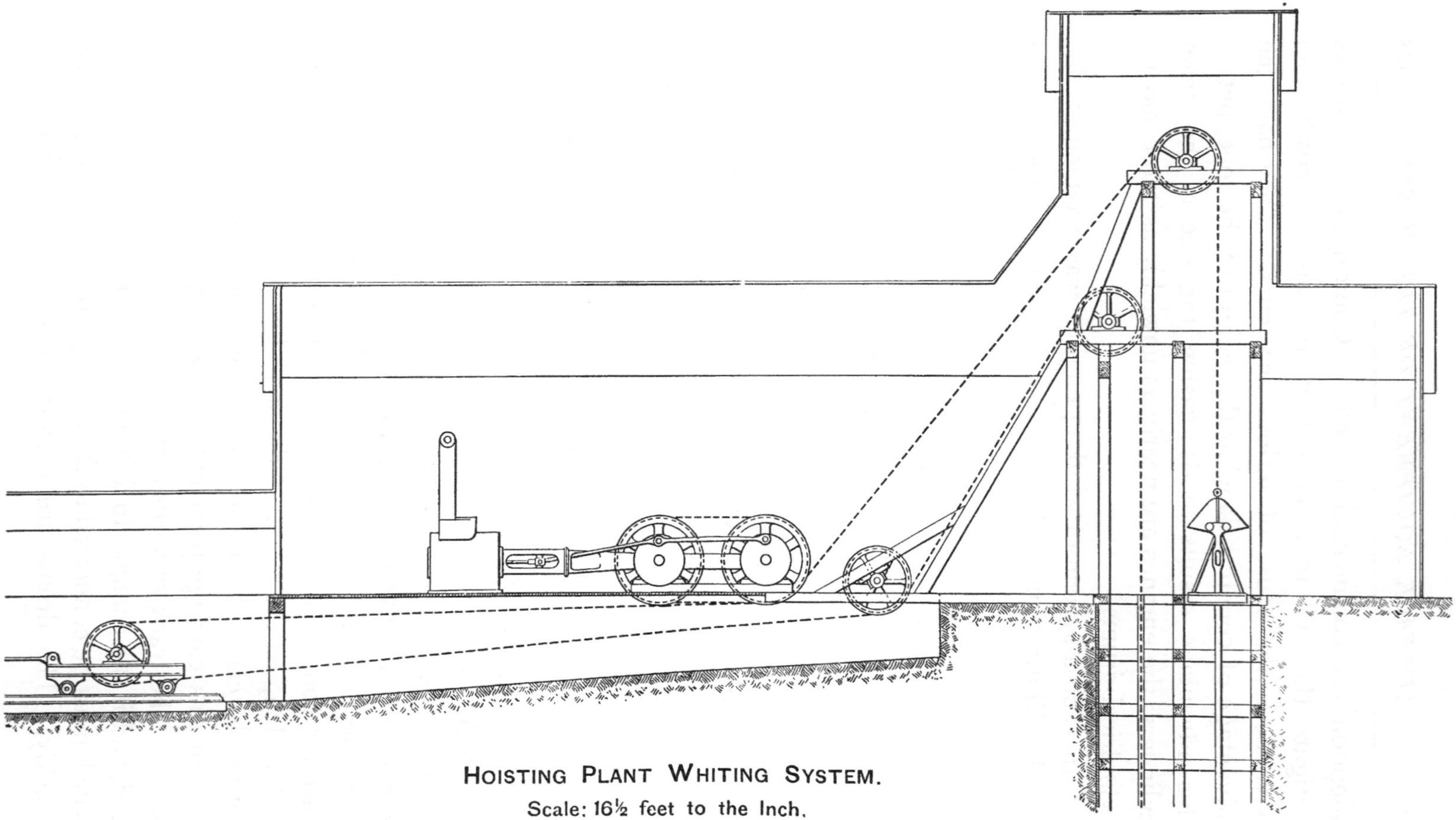

HOISTING PLANT WHITING SYSTEM.

Scale: 16½ feet to the Inch.

FIG. 72.

save wear on the ropes; but iron rings are now frequently used on a system known as Walker's patent. The latter, while providing a harder bed for the rope, do not wear; they also allow for any creeping of the rope on the drums without straining it.

Below we give a few particulars of a winding plant of this description as now used for sinking at the Calumet and Hecla Copper Mines in Michigan:—

Depth of shaft (present)	4426 feet
" (ultimate intended)	5000 "
Diameter of rope	$1\frac{1}{4}$ inches
" hoisting drums on trod	7 feet
Number of turns of rope round the drums	3
Revolutions per minute	90
Speed of skip in shaft (feet per minute)	2000
($1\frac{1}{2}$ to 2 tons rock hoisted per trip).	

The engines are connected directly to one drum, and, by parallel rods, as in locomotives, to the forward drum, which is inclined to the other by the width of one groove. They are a pair of tandem compounds, with Corliss valves, each high pressure cylinder of 16 inches diameter, low pressure cylinders of 32 inches diameter, and all having a 48-inch stroke. The exhaust steam from the low pressure cylinders goes to a Worthington jib condenser and air pump, maintaining about 25 inches vacuum. A foot brake only is used, reversing being done by a small steam reversing gear. The drums being small, the engine gets up to full speed in three or four revolutions, and steam is carried nearly to the end of the run. This engine has been used entirely for sinking the shaft, and works with extreme sensitiveness and regularity. The shaft itself is 13 feet by 23 feet inside timbers, and is divided into six hoisting compartments. The speed attained in sinking has been quite phenomenal. In one month it was sunk 127 feet vertical, including timbers, while more than 1000 feet were sunk in one year. Rand drills are exclusively used, with compressed air at 60 to 65 lbs. pressure. At the present depth of over 4000 feet the rate of progress is somewhat slow, being now 70 to 80 feet per month, some time being lost in getting away the rock at so great a depth. The main winding engines at this shaft are already in position, and will be running in about a year. They are on a colossal scale, and are expected to work with an economy of fuel hitherto unattained. Their drums are also on the Whiting system, 18 feet in diameter in the trod. It is expected they will hoist a load of 10 tons at the rate of a mile per minute, which will be far and away faster than any hoisting now being done in any part of the world.

A still deeper shaft is now being sunk at the Tamarack mines near by, its ultimate depth being intended to reach 6000 feet. This shaft is to be served by horizontal Corliss engines direct on a double conical drum 36 feet in diameter.

Another system used on the Continent, known as the "Koepe," presents some features of extraordinary interest. The rope simply leads a little more than half-way round a single narrow hoisting drum lagged with wooden blocks, each end being attached to the respective cages. The writer recently visited a shaft thus equipped near Essen in Germany, the details of which are of great interest. The shaft is about 750 metres depth vertical, and cages carrying six tubs of coal are worked in compartments side by side. The winding drum is of 8 metres diameter, the greatest speed being about 20 to 25 revolutions per minute. The engine is a side-by-side compound, having a high pressure cylinder of about 40 inches diameter, low pressure on the other crank of about 70 inches diameter, and a stroke of 80 inches each. There is a large intermediate receiver between the cylinders in which the steam is usually at about 30 lbs. pressure, while the pressure at the high pressure steam chest is about 90 lbs. per square inch. Its extreme simplicity seems remarkable, and it is being adopted at many shafts of great depth which formerly were supplied with the ordinary cylindrical drums. The writer failed to detect any slipping at either high or low speeds, and the driver said none had ever occurred.

Judging from these and many other examples of deep winding plants, there does not appear to be any difficulty, as far as machinery is concerned, in mining to depths of 5000 or 6000 feet. Differences of opinion exist as to whether it is best to wind with one pair of engines from shafts which are partly vertical and partly on the underlay ; but considering the objection to carrying steam pipes underground, and the expense of handling twice, it appears cheaper and more satisfactory to hoist loads of 3 to 5 tons each by one pair of winding engines at a moderate speed than to put down double sets of plant for accomplishing the same result. The objection is often heard that it is unnecessary to seek for the highest commercial economy should it entail any complication of machinery; but when we consider that greater economy in working means the development of properties, now unworkable by reason of their depth, or on account of the low grade of their ore, we cannot afford to neglect any item leading to the cheapening of the whole process from mining to the final treatment of the slimes. As the depth becomes greater, the natives must be carried to and from their work underground instead of climbing down ladders, which, although it restricts the time available for hoisting the rock through the shafts, enables them to work harder while underground.

Looking at the whole question from a general point of view, it appears that existing types of machinery will be gradually followed by others more economical and suitable for deeper workings, warranting the assumption that working at greater depths will not entail greater expenditure than at present. A notable example of what has been done in this direction is to be found at one of the diamond mines in Kimberley. For a period of six months before the reconstruction of the whole machinery plant the total number of loads hoisted was 550,000, with an expenditure for fuel of £33,000 sterling. For a similar period after the reconstruction 775,000 loads were hoisted from a depth 160 feet greater, with a total fuel expenditure of £11,000. Part of this enormous decrease in the fuel bill was due to the use of coal averaging five-eighths the cost for equal heat values, the remainder being due to more economical machinery and methods.

CHAPTER VIII

THE METALLURGICAL TREATMENT OF THE ORE

I. Milling.

Owing to the simplicity of composition of the Rand ores, and the favourable conditions under which the gold exists in them, comparatively little difficulty is experienced in their metallurgical treatment. Oxides and sulphides of copper, arsenic, antimony, bismuth, etc., which frequently render gold ores refractory, are seldom if ever found in quantities sufficient to exert any detrimental influence. Iron pyrites is present in "blue rock" in quantities ranging from 1 to 4 per cent, but the ore, even when pyritic, can be made to yield a high percentage of its gold by amalgamation. In fact, in the laboratory, with fine grinding and careful handling, practically all the gold in a sample can be recovered by means of quicksilver. Apart from cyaniding, therefore, which, in a great measure owes its success on the Rand to the absence of refractory elements in the ore, there is little that is novel or even unusual in the metallurgy of the Rand conglomerates. The features of milling that are in any way peculiar are mechanical rather than purely metallurgical. It may be said, as a general rule, that a high extraction by amalgamation is not regarded as of the first importance on the Rand, a fact which is due to the universal applicability of the cyanide process to tailings. The tendency on these fields is to sacrifice the conditions that favour high battery extraction to the attainment of high stamp duty or crushing capacity. The degree to which such conditions should be sacrificed in order to increase the quantities dealt with, is of course a question of economics which must be considered in the light of profits ultimately to be secured. High stamp duty is not entirely compatible with the best amalgamation results, nor, under present conditions, is the best total extraction possible with a low mill extraction, for the higher the grade of the tailings leaving the battery, the higher, as a rule, will be the value of the final residues after cyaniding. For instance, let us take the average mill yield on the Rand as 64 per cent of the total gold in the ore, which is approximately the case. Thirty-six per cent remains in the pulp leaving the plates. Neglecting concentration for the sake of argument, about 10 per cent out of this 36 per cent runs off with the slimes, leaving 26 per cent, of which say, according to cyaniding results, 70 per cent is recovered, equal to 18 per cent of the gold in the original ore, *i.e.* the total extraction would be 82 per cent. Taking now a mill extraction of 70 per cent, which we are informed on the best authority is obtained at the Robinson and at the Jubilee, the tailings treated would contain say 20 per cent of the gold in the ore, and the extraction by cyanide at 70 per cent would amount to 14 per cent of the gold in the ore, *i.e.* the total extraction apart from concentration would equal 84 per cent. As a matter of fact, the total extraction at the Robinson is about 91 per cent, since, although the tailings' extraction is lower, 10 per cent is recovered in concentrates; but the above comparison is sufficient to illustrate the importance of good amalgamation. On the other hand, it must be borne in mind that with a high stamp duty and fairly high extraction a greater total output may be obtained, the interest on which would tend to counterbalance the simultaneous loss due to a lower extraction. Obviously the object to be attained is the determination of the exact point at which the conflicting factors are balanced to the greatest advantage.

The pulp resulting from battery crushing generally consists of about 90 per cent of water and 10 per cent of solid matter in particles of all sizes, from the mesh of the battery sieving downwards. The solids, on the average, may be taken to be composed as follows:—96 per cent of quartz, 1 per cent of argillaceous matter, and 3 per cent of iron pyrites, the gold constituting, of course, an infinitesimal percentage. By concentration, which is generally practised in some form, the bulk of the pyrites and some of the coarser sand are separated. The pulp is then regarded as divisible into sands, generally spoken of locally as "tailings,"[1] and "slimes." The latter consist of the very finely divided silica, clay, and pyrites. No exact definition can be given of what constitutes "slimes," since the proportion separated as such in any particular case depends very much on the conditions under which the pulp is dealt with after it leaves the battery. The semi-plastic nature of slimes is due to a great extent to the fine state of division of their component material, and not necessarily to any predominance of clayey matter, which, in reality, is present in very small proportions.

The gold is disseminated in the cementing material filling the interstices of the conglomerate, and frequently adheres to the surfaces of the pebbles. In "blue rock" it is intimately associated with the pyrites, from which, near the surface, it has become freed by oxidation. Much of the gold is in a fine state of division, especially in unoxidised ore, but there is always a fair proportion of particles varying from one-hundredth to one-fiftieth of an inch in linear dimensions. As previously stated, the gold is probably all uncombined, and is generally readily amalgamated.

[1] Strictly speaking, "tailings" include the whole of the crushed ore.

In our opinion the loss of "float-gold,"[1] or even of gold remaining in suspension, but not actually floating, in tailings' water is very slight. The gold that remains in suspension after two minutes in still water does not, as a rule, amount to more than one or two per cent of the total. Since, moreover, battery water is seldom allowed to run away, the bulk being retained in the dams for re-use, the absolute loss of very fine gold is probably inappreciable. It is noteworthy, in this connection, that slimes are generally poorer than the sands from which they have been separated.

The course of treatment to which the ore is subjected, according to the most approved practice, may be briefly outlined as follows:—

The ore, as it comes from the mines, is dumped on to grizzleys, which separate the fine stuff from the coarse. The former, passing through the grizzleys falls directly into a bin, while the latter runs down the grizzleys to a sorting-floor or table. The pay-rock that remains after sorting is delivered to rock-breakers for preliminary crushing. The product of the rock-breakers, falling into the ore-bins, is eventually trammed with the fines to the battery bins. From the latter the ore is fed by automatic feeders into the mortar boxes where it is wet crushed. With few exceptions, quicksilver is added in the mortar boxes, and the usual methods of extraction of gold by amalgamated copper plates are followed. The pulp on leaving the plates is, in some cases, led over blanket strakes, or on to concentrators, by which it is relieved of pyrites and black sands; in other cases *spitzluten* are used for the same purposes. About this stage of the treatment the pulp is generally elevated some 20 or 50 feet by means of pumps or bucket-wheels, in order to

[1] The term "float-gold," which is of course familiar to all mill-men, is applied to flakey particles of the metal which, owing to their large superficial area compared to their weight, do not overcome the surface tension of water, and consequently are unable to sink.

PLATE XIII.—HEADGEAR (INCLINED SHAFT) AND ENGINE-HOUSE, GELDENHUIS ESTATE MINE.

secure sufficient fall for economically carrying on subsequent operations.

The amalgam obtained from the mortar boxes and plates is subjected to the usual operations of cleaning and retorting, the resulting gold being melted into bars.

Concentrates are either treated by chlorination at customs works or cyanided on the spot.

There is very little variation from the above general scheme of ore treatment. Sorting of the ore is sometimes omitted, and the preliminary ore-breaking, now generally effected outside the battery, is still on some mines carried on in the batteries. Concentration is sometimes entirely omitted.

Mill Sites.—The steep slopes or hillsides which form the best sites for quartz-mills are, in the Witwatersrand district, rarely found in convenient situations; in fact, the only sites possessing fair natural advantages are those of the Geldenhuis Estate and the New Rietfontein. As a general rule, on the Rand little more is looked for in a mill site than a fall of a few feet in a hundred, and even this is not considered essential on the site itself if ground with a slight fall, which can be utilised for the settlement and banking up of the slimes, exists in the vicinity. In the early days batteries were erected as close to the available water-supply as possible, in order to avoid pumping water to any great distance, the greater evils of transporting the ore, and of lack of space for collecting tailings, being incurred through ignorance of the true issue. At the present day, among the first considerations in erecting a battery are proximity to the main shafts, the minimising of transport, and the concentration of plant generally.

General Design and Construction.—There is no essential difference in general design between the mills on the Rand and those erected in the Western States of America and other gold-mining districts. The principal desideratum in a well-designed mill is obtained, viz. the continuous, automatic progress of the ore from its delivery into the ore-bins to its arrival at the tailings' launder, at the expense of a somewhat heavy timber substructure in the back or ore-bin portion, and the trifling loss of power expended in elevating the ore. Banked sites are rare. The end elevation (Fig. 76), page 192, represents the typical arrangement of a back to back mill.[1] The stamps are invariably in batteries of five, and usually two 5-stamp batteries are erected together, a clear space being left between each set of ten stamps. In most cases each set of five stamps has a separate cam shaft and pulley. An exception to the usual practice in arranging batteries has been made at the Ferreira, where forty stamps are being erected with a clear space between each five stamps. There can be no doubt that there is a great advantage and convenience in having plenty of room about all parts of a mill.

The timber mostly employed in the erection of the batteries is pitch-pine. Karri, a heavy, close-grained Australian wood, has also been adopted in some cases for mortar-blocks and for sills. Oregon pine, which, though lighter, appears to be more durable for battery foundations than pitch-pine, has not been adopted on the Rand for this purpose. At the Geldenhuis Deep the mortar-box foundations consist of massive concrete blocks, instead of the usual wood piles, a departure which to the best of our knowledge is without precedent. One of the reasons for this innovation is the soft and yielding character of the ground on which the mill-site is located, and the consequent advisability of covering a large area with a solid and coherent foundation. Concrete has the advantage that rotting is impossible. Its superiority in other respects has yet to be demonstrated. Each set of ten stamps has a separate foundation, so that there is no likelihood of the concrete cracking through unequal subsidence. In large mills there is often considerable risk of subsidence, even

[1] This drawing, which is a cross-section of the 280-stamp mill ordered for the Simmer and Jack, we are allowed to reproduce by the courtesy of Mr. S. B. Connor, Mechanical Engineer to the Consolidated Gold Fields of South Africa.

when the foundation pits are dug to bed-rock, owing to the existence of softer layers or patches underlying an apparently firm bottom. For this reason mortar piles are generally laid on concrete foundations. At the City and Suburban the pile-pits were dug down to sandstone, a depth of about 12 feet. The bottom was then roughly levelled with stone rubble, over which was laid a bed of concrete 3 feet deep and 8 feet wide. On this foundation, the surface of which was levelled true, were set the mortar piles 13 feet in length. The piles for each box are fifteen in number, bolted together by horizontal bolts, endways and across. The walls of the foundation are built up with 2 feet of masonry, and all the spaces round the mortar blocks are tamped hard with tailings. The blocks are further held in position by longitudinal timbers of 8-inch square section abutting against them, and bolted to the line-sills. At the Wolhuter the mortar piles are 15 feet long, set on a concrete bed 18 inches thick. At the Simmer and Jack new mill the piles will be 15 feet long, the pile-pits being 8 or 9 feet deep, built up with 2 feet 9 inch walls of stone. The pits are to be laid on a 2-feet bed of concrete, and all interspaces will be tamped hard with sand and rubble. Staying timbers bolted to the line-sills, as in most batteries, will keep the upper portion of the piles in position. The piles will be composed of eight timbers, each about 14½ inches square. The piles and the sills will be of Karri wood.

When the bottom of a mortar block has been levelled it is usually covered with indiarubber ¼ inch thick, on which the box rests. The mortars are generally held down by eight 1½-inch bolts, four at the front and four at the back. The bolts are generally about 3 feet in length, and are held at their lower ends by cotters, in recesses cut into the wooden piles, and are tightened up by nuts screwed on their upper ends against the bottom flanges of the mortar.

Ore-Bins.—These are generally made of triangular transverse section, the floor sloping at an angle of from 40° to

50° towards the discharge shoots, but in some mills they are built of rectangular section, thus doubling the cubic capacity, and utilising waste space. With the latter the ore

FIG. 73.—SUSPENDED CHALLENGE ORE-FEEDER AND HOPPER. Rod and Rubber Buffer lying at bottom.

forms its own natural slope, and there is less wear and tear of the lining, while any extra shovelling that may be necessary when the ore-supply falls short, is of small moment when compared to the advantage of having a larger reserve for an emergency. The capacity of mill-bins is from 5 to 11 tons per stamp, or from 24 to 60 hours' supply, varying to some extent with the rates of crushing. When the preliminary breaking of the ore has been effected outside the battery, the ore-trucks are tipped directly into the bins. There are usually two trucks along the top floor, so that the trucks can be discharged on either side, and the ore properly distributed.

Ore Feeders.—Automatic feeders are used in all the mills. Various kinds of feeders have been tried, but the Challenge is usually considered to give the most regular feed, and though expensive in the first instance, it has been adopted in the majority of batteries. All the varieties of self-feeders in use are actuated by the fall of one or other of the stamps. The lifting tappet, or another, strikes in falling the top of a rod, which, through an arrangement of levers, ratchets, etc., imparts a motion to the tray from which the ore is fed into the mortar boxes. The jerk or movement given to the tray, and consequently the rate of feeding, can be adjusted at will. The rate of feeding is also regulated automatically, since no blow is given by the tappet unless the ore in the boxes is sufficiently low to let the stamp drop to the regulation extent.

Mortar-Boxes.—Most of the modifications introduced in the design of mortar boxes for Rand mills have been made with the object of increasing the duty of the stamps. Thus the tendencies of late have been to narrow the width of the boxes at the issue and to decrease the depth, either in the original casting, or by the use of false bottoms and of linements. The weight of boxes has been much increased in order to adapt them to the heavy stamps which have become the rule. The principal differences in other respects between the more recent mortar boxes and those of the Homestake type are—(1) In the height of the feed-opening, which is generally from 18 to 19 inches above the bottom of the mortar. (2) In the length of the boxes, which is from $50\frac{1}{2}$ to 53 inches, allowing of spaces between the footplates of the dies. The objections to having the dies fit exactly into the bottom of the mortar are—that it is much more difficult to keep the stamps plumb over the shoes than to bring the dies into adjustment when some play is allowed, and the liability of the dies to get jambed and cause loss of time in their removal.

The inside width of mortars, from back to front, at the issue, varies from $13\frac{1}{4}$ inches to $11\frac{3}{4}$ when heavy liners are

FIG. 74.—MORTAR-BOX in use at the New Crœsus.

inserted. The depth from the lip is generally 9 inches or 10 inches, but with these depths it is impossible to get a very low discharge unless false bottoms are put in. With the usual 7-inch dies false bottoms must be used in any case in order to maintain the discharge at 4 or 5 inches, which is the customary height. In some of the later batteries, as at the City and Suburban and the Robinson new 40 heads, heavy cast iron liners, $\frac{3}{4}$-inch thick, are inserted round the whole interior of the lower portion of the box; these being replaceable will add considerably to the life of the boxes. The sand resulting from the crushing of banket seems to have a more than usually quick-wearing action.

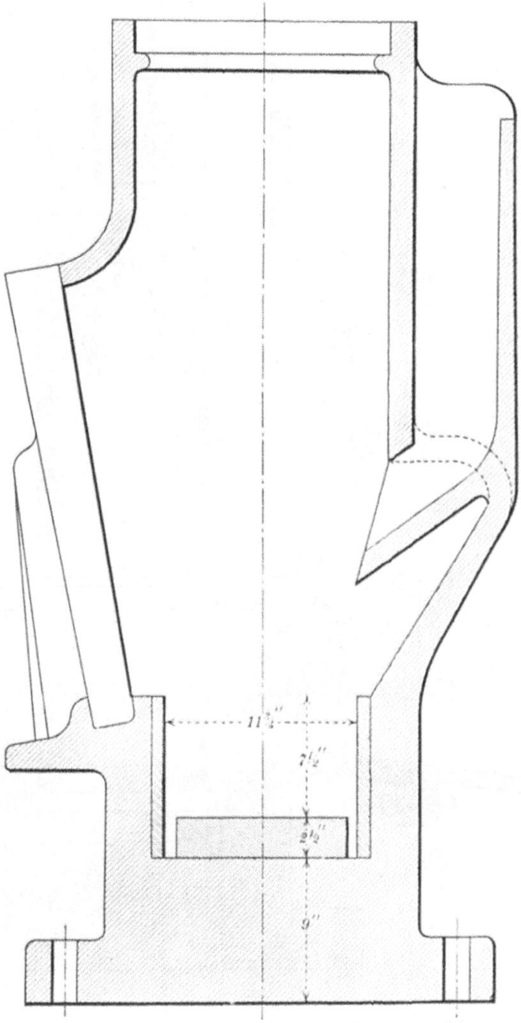

FIG. 75.—MORTAR-BOX at the City and Suburban.

Fig. 75 is a section of the mortars put in for the City and Suburban 160-stamp mill and for the Robinson new 40 head. Fig. 76 is the mortar introduced by Mr S. B. Connor at the Crœsus. This design, which was originally adopted at the

Alaska Treadwell, will also be used in the Simmer and Jack 280-stamp mill. The chief peculiarity of these mortars lies in

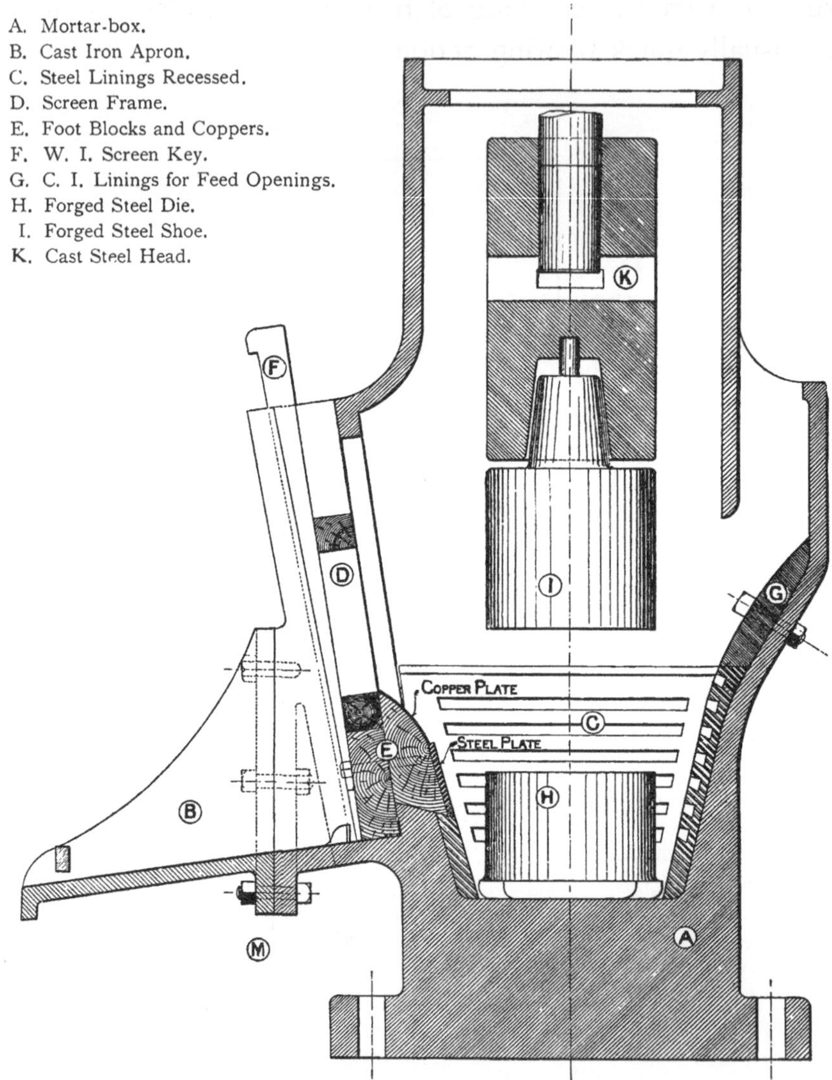

FIG. 76.—Section of the Mortar used at New Crœsus Mine, showing Stamp Head, Shoe, Die, and other parts.

the cast steel liners inserted at the ends and along the back. The slots or recesses in these liners, shown in the drawing, are for the collection of amalgam formed inside the boxes, and take the place of the usual copper plates. At the date of writing no clean-up has been made from any of these boxes on the Rand; but judging from the high extraction obtained at the Alaska Treadwell mill, the device is apparently of proved efficiency.

Dies.—The cylindrical portion of the dies used on the Rand is invariably 9 inches in diameter and 6 inches high; the total height of the die including the footplate is 7 inches. The footplate is generally from $9\frac{1}{2}$ to 10 inches square with bevelled corners. It is proposed at the Robinson to have dies 8 or 9 inches in height, but this alteration has not yet been made. The material is steel of special quality. Firth's dies and shoes are of compressed cast steel, Hatfield's are

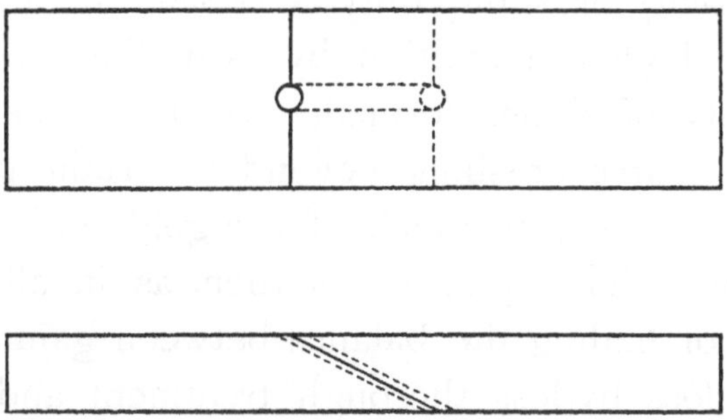

FIG. 77.—False Bottom for worn-down Dies.

manganese steel. The consumption is from $\frac{1}{4}$ to $\frac{1}{3}$ lb. per ton crushed.

As dies wear down, a false bottom of cast iron is usually inserted in the mortar, in order to bring the surface of the die nearer the lip, and maintain a sufficiently low discharge. These false bottoms are usually cast in two pieces, the best form being that shown in Fig. 77. In some cases two or three different thicknesses are kept, so that greater uniformity of discharge can be secured.

Battery Screen.—Woven iron or steel wire screens are used throughout the Rand. The size of hole varies greatly on account both of the variations in the number of holes to a given area, and on account of differences in the thickness of the wire with which the gauze is woven. Screening with only 100 holes to the square inch will be used at Buffelsdoorn in the Klerksdorp district, while in one or two cases on the Rand, as at the Chimes, 1200-mesh screening has been used. The majority of the Main Reef Mines, however, use screens with from 600 to 900 meshes to the square inch, in which the total area of mesh is from 40 to 50 per cent of that of the screening, the actual area of each hole varying from .0004 to .0007 of a square inch. A tendency exists, and will probably continue until some lower limit is arrived at, to adopt coarser screening. It is considered by many that, within limits, coarser crushing is advantageous, not only because the crushing capacity or stamp duty is raised, but also as being conducive to higher extraction by cyaniding, owing to the lesser quantity of slimes formed. At the same time, it is evident that coarse crushing beyond a certain point would fail to expose or liberate much of the gold, and thus lead to low extraction. The question is then, as in all matters of milling, one of finding the balance between gain by quantity treated and loss by less thorough treatment and extraction. At present, although undoubtedly the best mill extractions are made when finer mesh screening is used, the average of the mills obtain about 64 per cent, which may be said to be good, and perhaps the adoption of a somewhat coarser screen would not make a proportional difference. At the Buffelsdoorn mine, where the rock mills 4 or 5 dwts. to the ton, it was found this yield was affected to the extent of only ½ dwt. by the substitution of 400-mesh for 900-mesh screens.

Discharge.—The height of discharge in batteries is regulated, as the dies wear down, by the use of chuck blocks, or of sets of sieve frames with different widths of the lower side, and also by the use of false bottoms. In no batteries is the discharge allowed to become higher than 6 inches. The object in most mills is to keep it below 4 inches.

Stamps.—The tendency for the last three years on the

Rand has been to greatly increase the weight of stamps. From the 700 or 750 lbs. per head originally in vogue, the weight has in later specifications grown to 1000 lbs. and upwards, while in one instance, the Modderfontein 40-stamps battery, it will be 1250 lbs., probably the heaviest on record in a Californian stamp mill. No batteries are now erected on the Rand with stamps of less than 950 lbs. each, a fact which has, of course, an important bearing on the crushing capacity of the mills.[1] Weight has been added to all parts of the stamps except, perhaps, the tappets, which are seldom more than 130 lbs. Stems have been increased from 3 inches and 3¼ inches to 3½ inches in diameter, and they are frequently 15 feet in length. Heads generally vary from 17 inches to 23 inches in height, the diameter being 8¾ or 9 inches. Shoes are seldom less than 10 inches in height and 9 inches in diameter, and shoes 12 inches high are used in many mills. A method of equalising the weight of stamps on the wearing down of the shoes, which we believe was originally introduced on the Rand by Mr. Harland of the Robinson, has been very generally adopted. As the shoe—weighing, say, 200 or 230 lbs.—constitutes a very considerable portion of the total weight of the stamp, it follows that when it has worn down to half its original size the stamp loses weight accordingly. To compensate this to some extent several sizes of heads are used in one mill, and new shoes are put on to the lighter heads, the heavier heads taking the partially worn shoes. In this way much greater uniformity is maintained in the weight of the stamps throughout a battery than formerly. The "chuck shoe," an invention of Mr. C. Raleigh, is designed to serve the same end. Briefly described it has the shank of a shoe and the bottom of a head. The shank is of the dimensions of an ordinary shoe, and is fastened to a head in the usual manner; the bottom

[1] The average weight of the stamps in use last year in Witwatersrand mills was 850 lbs.—*State Mining Engineer's Report for* 1894.

recess of the "chuck shoe," then, holds a new or partly-worn shoe as a head would. It has the advantage of preventing the wear of the head itself. Fig. 78 is an illustration of this shoe.

From stamps weighing from 800 to 1100 lbs. there are the following variations in the weights of the parts:—

Tappets from 106 to 130 lbs. Heads from 260 to 365 lbs. Shoes from 180 to 230 lbs. Stems from 350 to 475 lbs.

Shoes 10 inches in height last from $2\frac{1}{2}$ to $3\frac{1}{2}$ months, the

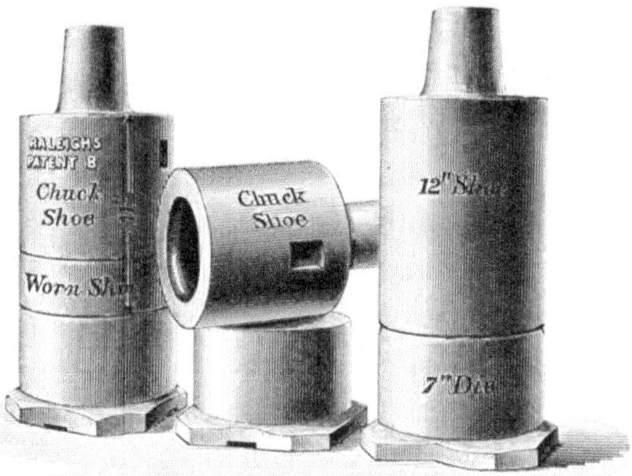

FIG. 78.—"CHUCK SHOE."

consumption averaging, when the rock is undecomposed and "blue," about $\frac{1}{2}$ lb. per ton crushed. A number of makes of shoes are used. Among the best are Firth's steel, Sandycroft, Fraser and Chalmers, and Hatfield's manganese steel. There is but little difference between the quality of these makes.

Cams vary somewhat in the length of sweep, in the weight and in the width of the face, according to the maker, and the weight of the stamps used. Several inventions with the object of overcoming the mechanical defects of the ordinary cam have been tried on the Rand. Of these the "Bally cam" combines the right and left hand cam, so that each stamp is lifted alternately on one side and the other, and in each battery of five stamps a right and a left hand cam can be engaged at the same time, so that the side thrust on the cam-shaft bearing is equalised. Side thrusts of the stamp, however, are not actually avoided. In Hart's patent each cam face is divided and forked to include the stem, and the cam engages both sides of the tappet simultaneously. In order to affect the rotation of the stamp one limb of the forked

FIG. 79.—THE BLANTON CAM.

cam is made slightly longer than the other. Side thrusts are thus very much minimised, and almost all friction on guides and bearings avoided. In the Blanton cam, which is becoming very generally adopted on the Rand, there are no keys. The cam is fastened to the shaft by means of a circular wedge, which is prevented from slipping on the shaft by two short pins dropping into two holes bored in the cam shaft. This method of fastening conduces to very great saving of time and trouble in replacing broken cams and shafts, since a whole set of cams can be taken off a shaft in less than half an hour.

In the lighter batteries the cam shafts are usually $4\frac{3}{4}$ inches in diameter, but for more recent heavy stamps they are made $5\frac{1}{2}$ inches, and at the Crœsus they are 6 inches. Cam shafts are usually about 1 to $1\frac{1}{2}$ inches distant from the stems.

Drop and Speed.—The drop given to stamps varies from 7 inches to 9 inches, the general average being about 8 inches. To maintain a steady average of 9 inches throughout a battery running at over 90 drops a minute is almost impossible. An $8\frac{1}{2}$-inch drop, with a speed of from 95 to 100 a minute, may be considered the best conditions of running obtaining on the Rand. The speed is seldom allowed to get below 90, nor to exceed 100 drops a minute.

The favourite order of drop is 1, 3, 5, 2, 4, although 1, 5, 3, 2, 4 and 1, 4, 2, 5, 3 are commonly adopted. Perhaps the last is the best for amalgamation; the first gives a good regular splash, and is considered by some millmen to give the best crushing results.

Water.—The water used in crushing is from 4 to $6\frac{1}{2}$ gallons per stamp per minute, probably averaging about 2000 gallons per ton crushed.

Duty of Stamps.—It will have been observed that in matters such as weight of stamps, speed, screening, low discharge, and the narrowing of mortar boxes, the prevailing conditions are such as favour high crushing capacity. The quantity crushed per stamp per 24 hours is in very few instances under 3 tons. In the majority of cases it is over 4 tons, while in some instances it is over 5 tons. The average rate of crushing for the whole Rand is about $4\frac{1}{8}$ tons per stamp per day.

The tendency is undoubtedly still towards an increased duty, and we may safely predict that in a few years' time from 5 to 6 tons will be a common result.

Copper Plates.—There is very great variation in the number, size, and distribution of copper plates. Apron plates

PLATE XIV.—INTERIOR OF MILL, NEW PRIMROSE MINE.

are universally adopted. Back plates inside the mortar boxes are more common than front or "chuck" plates, but both are used. Lip and splash plates are also used, but not in the majority of mills. Apron plates are, as a rule, 10 or 12 feet long and 4 feet 6 inches wide, the usual inclination being $1\frac{1}{4}$ or $1\frac{1}{2}$ inch to the foot. In some mills, as at the Primrose, they are broken by one or two riffles, but the general practice is to have them continuous from end to end. As far as results go, riffles seem to make very little difference. In all the mills quicksilver traps are fixed at the lower ends of the apron plates. At the Primrose the lip plates are 12 inches wide, with a slope of about 1 in 5, and there are also splash plates 14 inches wide sloping towards the boxes. No chuck plates are used, but there are inside back plates 8 inches wide. The back plates in the other mills vary from 7 inches to 11 inches in width, extending along the whole length of the mortar. Where chuck plates are used, they are generally varied in the same battery from 3 inches to 6 inches in width, according to the degree to which the dies are worn down. In some mills, as at the United Main Reef, no inside plates are used.

The usual thickness of all copper plates is $\frac{1}{8}$ of an inch. This is rather light, in our opinion $\frac{3}{16}$ or $\frac{1}{4}$ inch copper might be adopted with advantage. Inside plates are liable, if too thin, to get excessively dented, while apron plate, especially if steamed frequently, will warp out of level. Silver-plated copper is seldom used.

Amalgamation.—The extraction obtained by amalgamation in batteries on the Main Reef Series varies from 55 to 70 per cent or more of the total gold in the ore. The average results from milling oxidised ore are, perhaps, slightly higher than where the ore is undecomposed and pyritic, but the difference is much less than might be anticipated, and almost as high extractions are in some cases achieved from pyritic as were obtained from the "free milling" banket.

Probably the greater skill of millmen on the Rand at this later date has a good deal to do with this fact.

Average pyritic banket cannot be considered in any way a difficult ore to mill. As already stated, it contains none of the more rebellious sulphides and oxides; the gold is on the whole clean and readily amalgamable, and the amount of "flouring" is small. No chemicals are used except a weak solution of cyanide of potassium in dressing plates.

It is sometimes claimed that a good mortar-box saving is incompatible with the character of the ore and with the design of mortar generally adopted, but while some allowance must be made for variations in these conditions, many failures on this score may be attributed in a great measure to indifferent milling, particularly to mistakes of judgment in the use of quicksilver, to the use of water in excess, and to over-feeding, and even to inattention to such details as uniformity in drop, and in level of dies, and in weight of stamps. The very low discharge, which is commonly adopted in order to increase the duty of the stamps, is also against good mortar-box amalgamation. In some mills no attempt is made to save gold inside the boxes, the outside plates being entirely relied upon. Where no inside plates are used, but quicksilver is fed into the boxes, the proportion saved in the boxes is, as a rule, only 4 or 5 per cent. Where there are only back plates and no chuck plates, this proportion is much less than when both are used. Under the former conditions the percentage of the mill gold that is retained in the boxes is seldom more than 20 per cent, while in the latter case it may be as much as 40 per cent, and is as a rule not less than 30 per cent if the amalgamators know their business.

Generally speaking, very little of the "free" gold escaping from the mortar boxes gets beyond the first two feet of the apron plates and tables, and it is not necessary to scrape the lower portion of the tables more than once, or at most twice a week. The upper plates of the tables, on the other hand, are never left more than two days, while in most mills they are scraped every twenty-four hours.

The Robinson Mill.—We are enabled to give the following notes on the milling at the Robinson, where close attention is paid to amalgamation, while the duty obtained in crushing is equal to that in most other mills on the Rand.

We have pleasure in expressing our appreciation of the courtesy extended to us by Mr. H. Harland, the mill manager, to whom we are indebted for most of the following particulars :—

The milling grade of the ore crushed at present is from 18 to 19 dwts. per ton. The rock-breakers are at the main shafts, so that the ore comes ready broken to the mill. The bulk of the rock leaving the breakers will pass a $2\frac{1}{2}$-inch ring.

There are running at present 70 stamps in 14 batteries of 5 stamps each erected in one row. Forty additional stamps are in course of erection. Challenge feeders are used throughout the mill.

The ore-bins are of triangular section, with a capacity equal to about forty-eight hours' supply. The mortar boxes[1] are 52 inches in length. The inside width at the lip is $13\frac{1}{4}$ inches, and the depth from the lip to bottom inside is 10 inches. In the mortar boxes for the new 40 heads the width will be decreased to $11\frac{3}{4}$ inches by the insertion of $\frac{3}{4}$-inch linements round the lower portion of the interior. The lower edge of the feed opening is 19 inches above the bottom, delivering against the shoes or on to the centre line of the dies. The length of the opening is 24 inches. The back plates, 12 inches wide, slope upward from the level of the issue under the cast-iron feed shoot. Chuck plates are used from 3 inches to 6 inches wide, according to the degree of wear in the dies, four sizes of chuck-block being kept.

The discharge is maintained as nearly as possible at 5

[1] Fig. 75 is a section of the mortars in the new 40-stamp mill.

inches. The battery screens are wire gauze, 900 holes to the square inch (see Table 5028). The inclination of the screens is about 10° from the vertical, from the lower edge upwards.

The stamps of the 70-head mill, originally 800 lbs. each, have been weighted up to an average of 930 lbs. by using longer stems (14 feet average), and heavier heads and shoes. The shoes are 12 inches high, exclusive of the neck, and 9 inches in diameter, weighing 229 lbs. Two sizes of heads are used, and half-worn shoes are transferred from the lighter to the heavier heads to equalise the weight of the stamps. The heavy heads at the Robinson weigh 365 lbs., and are 23 inches in height, the lighter heads weigh 280 lbs., and are about 19 inches high.

The wear of dies is compensated by the use of false bottoms as well as by lower chuck-blocks. The false bottoms are of cast iron $3\frac{1}{2}$ inches thick, fitting into the bottom of the mortars.

The shoes are worn completely out, a discarded shoe weighing usually under 30 lbs. The life of 12-inch shoes is from $3\frac{1}{2}$ months to 4 months, the consumption being about 0.45 lb. per ton crushed. Dies last from 3 to $3\frac{1}{4}$ months, the consumption being about 0.3 lb. per ton crushed.

The stems in the 70-stamp mill are $3\frac{1}{4}$ inches in diameter, in the new 40 heads $3\frac{1}{2}$ inches.

The Blanton cam, described elsewhere, is used. The cam shafts are $5\frac{1}{2}$ inches in diameter.

There are two orders of drop in the battery, but 1, 4, 2, 5, 3, is preferred. The height of drop averages $8\frac{1}{2}$ inches, the speed being 97 blows a minute. The feed is kept very low, so low that the stamp has a slight rebound in striking. It is considered that the advantage thus gained in the duty of the stamps is important, while the breakages of stems, owing to the greater vibration, are no more frequent than in the generality of mills.

Screening lasts from 2 to 4 days.

Neither lip nor splash plates are used. The apron plates or tables are 12 feet long and 54 inches wide. There are no riffles or drops in the tables. The thickness of all the copper plates is about $\frac{1}{8}$ inch. A quicksilver trap is fixed at the end of each apron plate.

Quicksilver is added in the boxes with greater or less frequency, according to the grade of the rock, which varies considerably from day to day. Sometimes it is added at intervals of 10 minutes, never less than once an hour, and generally less than once in half an hour. As little quicksilver as possible is put into the outside plates.

The inside plates are scraped once every 3 to 6 days, according to the richness of the rock milled. The apron plates are dressed about every 4 hours, but amalgam is removed only every other morning. When the plates are dressed all lingering or adhering black sand and pyrites are collected, to be ground with quicksilver in a pan or batea. Much coarse gold is caught in this way, the saving amounting to a considerable percentage. A further saving by amalgamation is effected on the copper plates fixed on the distributors of the Frue vanners. The loss of quicksilver is about $\frac{1}{3}$ ounce per ounce of gold saved.

The total gold saved by amalgam amounts to about 70 per cent of that in the ore fed into the mortars, and the duty per stamp is 4.6 tons per 24 hours,—combined results which bespeak excellence of the mill administration.

The total extraction at the Robinson is about 90$\frac{3}{4}$ per cent.

Battery Staffs.—In large mills there is always a battery manager who takes no shift. The number of amalgamators employed is very variable, two men on each shift attending to from 30 to as many as 80 stamps. Besides amalgamators there are mechanics (from 1 to 4, including carpenter and fitter), general hands, and engine-drivers, and stokers. Natives

TABLE

MILLING

Name of Mine.	Number of Stamps.	Weight of Stamps. Lbs.	Inside Dimensions of Mortar Boxes.			Battery Screen.		Height of Discharge. Top of Die to lower edge of Screen. Inches.	Average Drop of Stamps. Inches.	Average Speed Drops per min.
			Length. Inches.	Depth from Lip. Inches.	Width at Issue. Inches.	Holes per square inch.	Size of each Hole in square inch.			
Crown Reef . . .	120	995	50½	9	13¼	700	.00061	6	8	95
Robinson . . .	70	930	52	10	13¼	900	.00044	5	8½	97
,, . . .	40	1100	51¾	10	11¾	...	...	...	...	...
Ferreira	40	850	50	9½	...	800	.00050	3½ to 5	8½	94
,, ...	40	1064	...	...	...	...	...	...	...	...
Jubilee	40	950	51	...	...	900	.00044	3½ to 5	8	90
City and Suburban .	80	956	50¼	10	11¾	700	.00061	6	8	96
,, ,, .	50	750	52	8	13	400	.00100	2½	7½	100
New Heriot . . .	60	980	...	...	...	700	.00047	3½ to 5	7½	96
Geldenhuis Estate .	80	950	...	...	...	} 700	.00057	2½ to 4	8½	96
,, .	40	1050	...	...	...					
Simmer and Jack .	100	800	...	...	...	900	.00044	3	7	92
New Primrose . .	100	1000	51	...	...	700	.00047	3 to 5	7½	90
,, . .	60	1100	53	9	12½	700	.00047	3 to 5	8½	90
May Consolidated .	50	800	...	...	...	700	...	3½ to 5	7½	95
,, .	30	1150	...	...	...	900	...	3½ to 5	8	95

5028.

STATISTICS.

Tons crushed per Stamp per 24 hours.	Inside Plates. Dimensions.		Outside Plates. Dimensions.				Average grade of Ore. Dwts. per ton.	Per-centage caught by Amalgama-tion.	Remarks.
	Back Plates. Width in inches.	Chuck Plates. Width in inches.	Lip Plates. Width in inches.	Splash Plates. Width in inches.	Apron Plates. Width. Inches.	Apron Plates. Length. Feet.			
5	9	6	None	None	54	12	8	57.74	Caught inside mortar 34.9 % of 57.75.
4.6	12	3, 4, 5, & 6	None	None	54	12	18½	...	...
...	...	...	...	...	...	...	...	...	...
3.75	11	3¼	None	None	...	...	21	60.0	...
...	...	...	...	...	...	...	...	...	...
3.25	9	None	None	None	...	...	10½	70.0	Caught in mortars 15% of 70.
4.75	9½	6, 4½, & 3	None	None	58	12	8	54.7	Caught in mortars 22½%.
3.75	5	None	12	7	56	10	7	56.4	Caught in mortars 17½%.
4.3	None	3½ to 6	None	None	...	...	9¼	62	...
4.25	11	None	None	None	54	15 *	7	64	...
3.9	None	None	...	...	...	...	9¾	...	Caught in mortars 4½% of mill gold.
} 5.25 {	7	None	...	...	54	11	} 6¾	61	Caught in mortars 14% of 61.
	12	None	12	14	54	11			
4.3	None	None	None	None	...	...	} 7¼	66	{ Caught in mortars about 10% of mill gold.
5.5	9 + 7	None	None	None	...	...			

* There is a drop of 6 inches from the upper 10 feet of apron to the lower 5 feet, and a splash plate.

are employed on the ore-bins, shoots, etc., where necessary, and as general hands on work which does not require special skill. In small batteries the mill-manager or head amalgamator sometimes takes a shift in the day time.

At the Simmer and Jack 100-stamp mill the staff consists of a battery manager, 1 senior amalgamator, 6 amalgamators, 2 juniors, 2 men in the hoist, 2 general hands, 2 fitters, 1 carpenter. The total wages paid to the above hands per month are about £375. Engine-drivers, stokers, and natives, not included, probably add about £150 to the above.

At the New Primrose 160-stamp mill (back to back) there are a battery manager, 2 millmen (on day shift only), 3 amalgamators, 6 junior amalgamators, 1 battery mechanic, 1 carpenter, 1 timekeeper, 3 engine-drivers, 3 stokers, 24 natives in the mill, and 4 natives in the engine-room. The total wages paid to the above staff, including natives, amounts to about £600 per month.

Where vanners are used, of course, extra hands are necessary.

II. Concentration.

The pulp on leaving the plates is dealt with in various ways, but some degree of concentration, or at least of separation of the coarser sands and pyrites, is usual. For close concentration Frue vanners, two or three to every 5-stamp battery, are generally employed, the only other forms of concentrators that have been continuously run being the Scoular tables and gilt-edge concentrators at the Village Main Reef. Blanket strakes have in the past been pretty generally used; but though very well adapted to free milling banket, they save only a comparatively small percentage of pyrites, and latterly, even where concentrating machines are not used, they have been discarded, and replaced by *spitzluten*, *i.e.* hydraulic classifiers.

In considering the best method of dealing with the pulp

leaving the battery plates, there are, in the present state of advancement, practically three alternatives :—

1. To subject it to close concentration by a mechanical concentrator such as the Frue vanner, and then treat the tailings by cyanide after separation of the slimes, the concentrates being usually chlorinated.
2. To separate, by means of *spitzluten*, a larger percentage of coarse sands and pyrites, which can be charged on the spot into cyaniding vats for treatment, and then to treat the remaining tailings by cyanide after separation of the slimes.
3. To omit all concentration, and treat the tailings by cyanide after separation of the slimes.

The Frue Vanner. — The essential part of the Frue vanner is a slowly travelling rubber belt, to which a quick lateral reciprocating motion, or side shake, is imparted. The belt has a slight inclination in the direction opposite to that in which it travels. The pulp is delivered on to the belt near its upper end, and the concentrates, being precipitated on to the rubber and adhering to it, are carried over the upper end. The lighter tailings are carried down with the water towards the lower end of the belt. The vanner is capable of very nice adjustment in respect of (1) the inclination of the belt, (2) the speed of the travel, (3) the rate of side shake, (4) the quantity of water added. Table 5029 gives the results obtained by vanner concentration at several mines on the Rand.

The advantages of close concentration by means of vanners or other machines are—

That a certain percentage of the gold in the pulp leaving the plates (roughly 20 to 50 per cent) may be retained in a small bulk of material, the treatment of which, though costing several pounds per ton of concentrates, adds little (one or two shillings) to the cost per original ton milled, while very little short of a cent per cent extraction can be obtained from such

TABLE, 5029.

PERFORMANCE OF FRUE VANNERS.

MINE.	Year.	Tons crushed during period.	Total Value of Ore.	Gold won by Amalgamation.	Tonnage of Concentrates caught.	Proportion of Concentrates to total Ore.	Gold in Concentrates per ton.	Gold caught in Concentrates per ton crushed.	Percentage of total Gold in Concentrates.	Percentage of Gold leaving plates in Concentrates.	Value of Tailings from Vanners.	Cost of Concentration per ton crushed.
			Dwts. p. ton	Dwts. p. ton	Tons.	Per cent.	Oz. dwts.	Dwts.	Per cent.		Dwts. p. ton	Pence.
Wemmer . . .	1894	27,640	17.1	10.7	628	2.5	5 6	2.4	14.05	37.5	4.0	...
,, . . .	1895	29,535	...	12.94	800	2.6	6 3	3.3	...	...	...	11½
Robinson . . .	1893	94,842	27.64	19.12	2678	2.8	4 0	2.25	9.11	30.0	5.8	...
Ferreira . . .	1893	24,736	27.49	14.8	918	3.71	6 14	5.89	21.41	46.0	6.75	8¾
Worcester . . .	1893	12,528	18.46	10.66	234	1.87	4 8	2.87	15.56	36.7	4.9	7½
,, . . .	1894	11,914	...	23.0	454	3.9	4 18	3.7	...	...	...	...
Chimes	1894	47,104	...	9.55	502	1.07	4 12	1.0	...	...	...	8
Langlaagte Estate . .	1893	107,257	...	5.75	700	0.65	3 7	0.43	...	...	...	4¾

concentrates by chlorination; also that any coarse gold saved in such concentrates can be more readily recovered by chlorination than it could be otherwise by cyanide treatment. On the other hand, as compared with direct treatment, or the use of spitzluten, are the cost of obtaining the concentrates, which is from 5d. to 11d. per ton milled, and the fact that even where the best concentration results are obtained the necessity for subsequent cyaniding of the tailings is not as a rule obviated. Of course the whole question is one of arriving at the difference between total extraction value and total cost by the respective courses pursued.

The question of the best method of treatment of pyritic tailings was reported on in May 1894 by a committee of investigation appointed by the Chamber of Mines. The conclusions arrived at by a majority of a sub-committee (one dissentient) at that time were as follows:—That concentration (*i.e.* close concentration) is usually a perfectly unnecessary process, as in nearly every case it has to be followed by a subsequent treatment of the tailings carrying at least 60 per cent of the gold (*i.e.* of the gold in the pulp leaving the battery). That in the case of low grade ores, like the Langlaagte Estate and Jumpers, although the concentrators show a profit, it is easy to show, from well-proved figures, that concentration (*i.e.* close concentration) is a failure. A letter was sent by Mr. J. R. Williams, then at the Langlaagte Estate, in support of this opinion. Although not definitely stated, there is a tacit admission in the report that close concentration may be advisable in the case of high grade ores, and under certain possible conditions of the ore and the gold, such as would render abortive attempts at economical cyaniding.

We have elicited the following expressions of opinion on this subject from metallurgists intimately associated with the treatment of Rand ores:—

From Mr. J. R. Williams of the Crown Reef—"that close

concentration is usually a perfectly unnecessary process, except in cases where, on experimental trial, the concentrates can be proved to contain gold unaffected by cyanide; or, should this class of ore decompose too much, then, in either case concentration becomes necessary, and further, in cases of concentrates very rich in gold, figures will prove when concentration becomes profitable."

Mr. Von Gernet, of the Rand Central Ore Reduction Works, after expressing practically the same opinion, adds: "With regard to concentration by Frue vanners, the method in use on the Rand of concentrating without previous classification is scientifically wrong, and gives crude results. Much better results would be obtained by submitting the pulp to a preliminary sizing, and then concentrating each class thus obtained on a separate vanner."

From Mr. Darling, of the Robinson:—"The question of close concentration depends to a very large extent on the original value of the ore milled. For the subsequent method of treatment—whether by chlorination or cyanide —it is generally admitted that a better result is obtained by having a certain proportion of sands mixed with the concentrates. A company selling concentrates to a customs works may gain by close concentration, as they are charged, according to tariff, so much per ton for treatment by chlorination. If the concentrates are very pyritic (closely concentrated), then in treatment it is found advantageous to mix with some other concentrates containing a large proportion of sands. This applies also to subsequent treatment by cyanide, so that when the ore is sufficiently rich to produce a good concentrate without *very* close concentration, and afterwards leave the tailings rich enough to give a product which will give a good profit on treatment by cyanide, the use of Frue vanners is advisable."

From Mr. Bettel, of the New Primrose:—"I do not regard close concentration of tailings from Rand ores an advantage for the following reasons :—"In deep level ores the finely divided gold is found in intimate mixture with the pyritic matter, part of which, being extremely friable, is reduced to such a fine state of division that, except by methods and appliances too costly to be introduced, the residual sands and slimes, after close concentration, will always pay for retreatment by cyanide. It must be borne in mind that the Frue vanners in use here are made to do nine-fifths the amount of work they do in America, where the gold is much coarser; even with this admission this class of machine will not save coarse sands to which adhere particles of gold, for, by reason of their specific gravity and size, these coarse sands are washed down with the tailings, which are enriched by the gold present in the mineralised sands. "It follows that close concentration, so-called (for a 21 per cent concentration is very inferior work), is a distinct disadvantage as a preliminary to cyanide work, for it leaves in the tailings coarse materials we desire to get out for prolonged treatment, and removes some gold-bearing material which is most easily treated by cyanide."

Undoubtedly the tendency throughout the fields is to omit close concentration in the treatment of ores of average or low grade. In considering the question at the present time it is necessary to take into account the almost daily decrease of difficulties obtaining in cyaniding generally and the further economies which will undoubtedly be achieved. The cyanide process is clearly progressive, while concentration and chlorination have during the past three or four years practically made no step forward on the Rand.

The use of spitzluten may be considered as an intermediate course which has the following advantages :—

1. The trifling cost of separating coarse and heavy material.

2. The comparative facility with which the product can be treated by cyanide and a good extraction obtained.

Chlorination.—Chlorination, which is the process best adapted to the treatment of high grade concentrates, is expensive unless conducted on a very large scale, so that few companies run a chlorination plant. Therefore, if a high extraction can be obtained by cyanide from coarse sands and pyrites (such as the 92 per cent claimed at the Crown Reef), the profit that would in most cases go to the customs works need not go out of the hands of the producing company, since a special cyanide plant can always be economically run on every mine.

With regard to the use of the chlorination process on the Rand, Mr. Bettel writes as follows:—"I am of opinion that the chlorination process is not a necessity here, as from the deepest levels yet reached the ore is perfectly amenable to classification and treatment by cyanide after amalgamation. Undoubtedly a small number of companies will continue to work their Frue vanners for some years to come, but as the tendency is to treat concentrates and tailings by cyanide, I do not think the chlorination process will extend; indeed, at present the chlorination companies (Robinson, Rand Central, Simmer and Jack, and Transvaal Chemical Company) have some difficulty now in keeping their furnaces going constantly."

And Mr. A. von Gernet:—"I think chlorination will become cheaper by reduction in costs of labour and fuel, and there is no doubt that it will always be used to a greater or less extent on the Rand, because there are certain classes of ore which by reason of their copper and other metallic contents are unsuitable for the cyanide process, and subsequently will have to be concentrated, and the concentrates treated by chlorination. The process is undoubtedly a good one on account of the uniformly high extraction obtained and the high standard of fineness of the gold produced."

Mr. Darling's experience goes to show that a concentrate up to about two ounces may be treated satisfactorily by cyanide, but over that amount it is preferable to treat by chlorination.

CHAPTER IX

METALLURGICAL TREATMENT OF ORE—*continued*

III. Treatment by Cyanide

The history of the cyanide process in South Africa[1] dates from the year 1889, when some mining men in the Barberton district, attracted by the results obtained by the MacArthur-Forrest process of gold extraction, formed the Gold Recovery Syndicate, for the purpose of negotiating with the Cassel Gold Extracting Company, the owners of the MacArthur-Forrest patents, for the right to work the process in South Africa. This right they in due course obtained. The Cassel Company undertook to make extensive and complete trials and demonstrations to prove the merits of the process and its commercial utility; and for this purpose they sent out a small plant from Glasgow, which was erected near the Salisbury Battery on the Rand. The purchase of concentrates and their treatment at these works was immediately commenced, the whole of the operations being open to the public.

The public demonstrations at the Salisbury plant were continued from June till August 1890, and in that time concentrates and blanketting from various mines were treated, proving that the process was not only adapted to

[1] We are indebted to W. W. Webster, Esq., for some interesting notes on the introduction of the MacArthur-Forrest process in South Africa, of which we have made use above.

what is usually termed free milling ores, but also to the treatment of pyritic ores. The following is a list of the ores treated, with the results obtained:[1]—

Ores.	Quantity Treated.	Assay Value per Ton.			Fine Gold Extracted.
	Lbs.	Oz.	dwts.	grs.	Per cent.
Free Milling Concentrates from the Robinson G. M. Company	2,744	5	18	9	89.6
Concentrates from the Salisbury Company	18,802	6	3	11	93.2
Residues from Pan Amalgamation from the Salisbury Company	22,400	2	7	8	80.2
Tailings from City and Suburban	13,440	0	17	4	85.0
Tailings from the Langlaagte Estate	2,176	6	13	2	93.9
Pyritous Ore from the Ginsberg Syndicate	5,670	1	7	0	90.0
Pyritous Ores from the Nooitgedacht Company, Klerksdorp	6,210	0	19	14	90.8
Pyritous Ores from the Wilkinson Company, Klerksdorp	1,763	0	8	4	80.0

By November 1890 the demonstration experiments carried out under the supervision of Mr. Alfred James at the Salisbury Gold Mining Company were practically finished,[2] and the more serious question of treating tailings by this method on a commercial scale arose. It was decided by the Gold Recovery Syndicate to erect a plant on the Robinson Gold Mining Company's property, and Mr. Darling, who had taken part in the experimental work at the Salisbury, was selected to take charge.

On the 23rd November 1890 the works at the Robinson Gold Mining Company were started,[3] the Gold Recovery Syndicate having entered into a contract to treat 10,000 tons of Robinson tailings on a sliding scale according to the assay value, and it may be of interest, as showing how little was known of the process, to state that tailings assaying

[1] As published in the *Star* of 16th August 1890.

[2] Shortly thereafter the Salisbury Company adopted the process under a royalty payment, and erected a small plant for the treatment of their tailings alongside the Cassel Company's plant. This company was therefore the very first company on the Witwatersrand gold-fields to adopt and use on its own account the MacArthur-Forrest process.

[3] Mr. Darling has supplied us with the notes in the working of the process at the first Robinson Cyanide Works.

8 dwts. or under were not to be paid for. The Robinson Company had the option of taking over the works at the end of the contract, if satisfied with the results; and a representative of the company had free access to the building at any time to take samples of the residues and tailings. Instructions were received to put through a certain tonnage without regard to the extraction, as it was arranged that 1000 tons had to be treated by a certain date, and this order, taken together with the fact of having to work with an incomplete and leaky plant, was not conducive to getting the best results.

The theoretical extraction for the first month's working (January) showed 82 per cent, but the actual gold obtained was as disappointing as has proved to be the case in starting other cyanide works since that time. In February, however, the results were much more encouraging, the theoretical extraction showing 80 per cent, and the actual gold recovered being 79.4 per cent of the assay value of the tailings. During this month weekly clean-ups were made, and the results cabled as soon as possible to London, where the Gold Recovery Company was in course of flotation. The total output for February from 1447 tons (of 2240 lbs.) treated, amounted to 1547 ounces 6 dwts., of an approximate value of £4700. On this return being cabled to England the African Gold Recovery Company was floated. As at this time there was some question as to the comparative value of cyanide and mill gold, the return was calculated and reported as equal to 1285 ounces of Robinson mill gold.

During the six months these works were run by the Gold Recovery Syndicate a total tonnage of 10,485 tons (of 2240 lbs.) were treated, yielding over 8000 ounces of bullion, or equal to over 6000 ounces of fine gold, the approximate value being £24,500. The theoretical extraction for that period was rather over 80 per cent, and the actual extraction as near as could be ascertained 75 per cent. Considering that the process was entirely new, and the plant not as satisfactory as it might have been, the results were sufficiently good to justify the Robinson Company in taking advantage of the contract with the Gold Recovery Syndicate to take over the plant on the completion of the contract for 10,000 tons.

In 1889 tailings might be considered as at a discount, there being no process at work on the Rand that could profitably treat them; in fact, it is known that some companies were anxious to get rid of them, at any rate taking no pains to conserve them; and where the nature of the ground permitted they were allowed to flow away, and in many instances these derelict tailings have since been pegged out by others and turned to profitable account.

The year 1890 saw the introduction of the cyanide process into this district. From the demonstration trials in which only comparatively small quantities of concentrates, blankettings and tailings were tested, a return of about 300 ounces of gold was made. The cyanide output on the Rand reached, in 1891, close upon 35,000 ounces.

In 1892	it was	160,168	ounces, valued at	£502,408.
" 1893	"	304,498	" "	£938,870.
" 1894	"	587,388	" "	£1,772,472.

The great bugbear amongst mining people in the earlier stages of the history of the Witwatersrand was the knowledge, that as soon as the free milling ore was exhausted the pyritic nature of the banket would increase the difficulties in the extraction of the gold; hence the greatest anxiety was felt as to the probabilities of a profitable method of treatment being discovered. To those who pinned their faith to deep level working the question of the profitable working of the ore when they got it to the surface was a most difficult one. The difficulty was solved when it had been proved that the pyritic ore was amenable to cyanide treatment, and increased confidence in the future of these fields was the inevitable result. There is no doubt that the industry owes much to those who introduced the cyanide process on the Rand, and have since brought it to its present stage of perfection. Its successful introduction was undoubtedly due, as we have shown above, to the enterprise of the African Gold Recovery Company, while the elaboration and perfecting of the process has resulted mainly from the efforts of the Central Ore Reduction Company in the person of Mr. Charles Butters, and of individual workers such as Mr. W. Bettel and Mr. J. R. Williams.

A glance at the curve showing the tonnage treated, given on Production Chart (see Appendix), will give the best idea of the rapid strides with which the treatment by cyanide has advanced. At first most of the companies had large accumulations of tailings to deal with, and the plants were designed with a capacity for treating a considerably larger quantity than the current production, but the falling off in cyanide production that might have been expected from the exhaustion of old accumulations has been more than compensated by the number of new works started and the increased crushing power of many of the companies. At the present day the pulp leaving the battery is led directly into the precipitating vats or tanks, and the sands thus obtained are subjected at once to the cyanide treatment. There are two principal methods in use for collecting the sands for immediate treatment, they are—

1. Direct filling, first introduced by Mr. H. Jennings at the Heriot, City and Suburban, Crown Reef, and Paarl Central Gold Mining Companies.
2. Intermediate filling, introduced by Mr. Charles Butters, and in use at the works affiliated with the Central Ore Reduction Company.

In the first method the pulp from the battery passes into a hydraulic separator, a form of spitzlutte, by means of which the slimes and fine sands are allowed to flow away to the slime pits, while the coarser sands are conveyed to settling tanks and deposited by means of an indiarubber hose, which is kept moving about by Kaffirs in order to effect an even distribution of the sands being deposited.

The second method, or intermediate filling, consists in delivering the pulp by a mechanical distributor, on the principle of the Barker's mill, into a tank kept full of water, in which the sands are deposited, while the slimes escape with the waste water over the rim of the vat into an annular launder. Once deposited the sands can, according to their nature, be treated directly, or they may be removed by bottom discharge to the leaching vats, or they may be submitted to a preliminary treatment with cyanide in a collecting vat, and then removed for a second and final treatment in the leaching vats. The latter method constitutes the so-called double treatment.[1] The gold solution obtained by leaching is conveyed from the vats to the precipitating boxes, and the gold is there precipitated by passing the solution over zinc shavings (MacArthur-Forrest process), or by electrolysis on to lead foil (Siemens-Halske process).

The above is an outline of the cyanide process. We do not propose here to discuss the details of the industry, as the subject is a big one, and much has already been written on it.[1] There are, however, several points in regard to the methods of treating the mill products which are of the greatest importance to the industry, and which at the present day are being carefully studied by the leading metallurgists of the Rand. We refer to the questions concerning classifying, direct or intermediate filling, single or double treatment, slimes treatment, and the cyaniding of dry crushed ore. In order to obtain the latest phases of development in the cyanide industry,

[1] For a valuable account of the Chemistry of the Cyanide Process, see W. R. Feldtmann's *Notes on Gold Extraction by Means of Cyanide of Potassium*, published in Johannesburg, 1894.

we framed a series of questions touching on these points, and submitted them to the best authorities for consideration.

The questions on which an opinion was requested were the following:—

1. The best method of handling the mill products.
2. The best process for precipitating the gold solutions.
3. The question of slimes treatment.
4. The practicability of the direct treatment of Rand ore by cyanide.

1. *The Handling and Treatment of the Mill Products.*

On this subject Mr. Charles Butters has favoured us with the following communication:—

"The mechanical preparation of the pulp for the cyanide treatment is a very much more important operation than it is ordinarily considered. This factor has been brought to prominence, as have all the other factors, in relation to the cyanide process, by reason of the enormous size of the industry on these fields, and the losses of gold, unaccountable in many cases, which have been brought out in the working of the process. In the early days of the industry the pulp was run directly into retaining dams. At the heads of these dams a sort of natural classification or separation took place. About 50 per cent of the pulp so deposited was workable, the other 50 per cent was badly mixed with slimes. If we take the mill records we find that not much over 50 per cent of the rock crushed was afterwards put through the cyanide works. Some of the pulp ran to waste, some was blown away by the wind, and some was mixed up with slimes. The average gold recovery was about 65 per cent, or in other words, 65 per cent of 50 per cent, *i.e.* 32½ per cent of the gold run into the dams was all that was recovered by cyanide treatment. Two serious attempts were made in different directions, both about the same time, to obtain a larger proportion of this gold, by retaining at once a larger percentage of workable pulp. The one method was direct treatment, introduced by Mr. Henning Jennings, in which the total pulp from the battery, instead of being allowed to flow into retaining dams, was passed through spitzluten, by means of which a certain amount of the slimes was eliminated, the balance of the pulp running from the bottom of the spitzluten into a leaching tank for direct treatment by cyanide. This method is a very great improvement upon the ordinary system of running into a dam and removing the pulp thence into the vats. A great many obstacles have been encountered and skilfully overcome by Mr. Jennings in this treatment. This method of collecting sands ready for treatment is theoretically correct, and absolutely the cheapest method, as far as treatment costs are concerned. In practice, however, difficulties crop up which at first sight are not apparent. Let us stop for one moment and ascertain what is the result we wish to obtain. This, undoubtedly, is the most perfect extraction of the gold obtainable at the least possible cost. Now, in order to get the highest possible extraction, we must have complete solution of all the exposed gold ; and this can only be obtained by permitting the cyanide to come into contact with the entire surface of each grain of sand. It is obvious that in order to accomplish this result each particle of sand must be washed clean from clay ; the particles must be as nearly as possible of the same size, in order to have the greatest number of interstices, and to accomplish this, since the grains of sands are of many different sizes, we must have the various sizes more or less assorted for treatment in different tanks. To arrive at this result we must classify or sort the sand ; and when this is thoroughly and completely done, any liquid poured upon these sorted sizes runs through them as it would through a sieve. Air also passes readily through them, and this means rapid solution of the gold by the aid of the oxygen, and there is a complete and perfect contact of the cyanide solution with the gold exposed by crushing. On such products an extraction of from 80 to 90 per cent may be easily and readily obtained. By the direct treatment process a perfect extraction of the gold may be looked for at once without double treatment, or without removing from one tank to another. Such pulp weighs less per cubic foot on account of the interstices than unsized pulp. For instance, at the Metropolitan Mine we find that 26 cubic feet of unsized pulp, after transferring from one vat to another, were required to make a ton, while after sizing 29 cubic feet went to the ton, showing that there were actually 3 cubic feet more of interstices in the sized pulp than there were in the unsized pulp. Now, if by the direct treatment we can save the cost of a second handling, and at the same time get as high an extraction, we have obtained the theoretical result. This result, however, has not yet been achieved by Mr. Jennings' direct treatment, because up to the present there has been no clear conception of what was required to be done. I have no doubt that this ideal result will ultimately be obtained by this method. The perfection of the present system lies in the direction of a multiplication of spitzluten, with more perfect washing in the spitzluten than at present, and the laying down in the direct treatment tanks of different sizes of absolutely clean washed sands, free from even one per cent of slimes, which means the construction of very perfect and complete washing plants. When such sands have been obtained I do not think that double treatment will be found necessary. The pulp which is now retained by the direct treatment tanks contains more or less slime, forming impervious layers, which prevent complete contact between the sands and the cyanide solution, and also excludes free access of air; hence the oftener this pulp is turned over and retreated up to a certain limit the more perfect is the extraction. This impervious condition of the sands has called into existence the system now known as double treatment, which, it is readily seen, would not be required if the sands were completely sized.

"The one other method that has been adopted for the collection of sands is known as the Butters and Mein's collecting vats. In this system the pulp is run on to a mechanical distributor, and is automatically distributed into a vat filled to overflow with water. By raining the sands down upon a water surface the slimes overflow constantly into an annular launder surrounding the vat. This machine acts in a way as a sizing machine by retaining nearly all the sands, fine and coarse, and allowing only the slimes to overflow. If the diameter of the vat is proportioned to the character of the pulp coming into it, a very clean product is obtained. If the vat is too large the slimes settle; if the vat is too small, fine sands overflow. The chief difficulty in this system is to obtain means to adjust the nature of the pulp to be obtained. The diameter of the vat is fixed, the quantity of the overflow from the battery is also fixed. If half a battery should be hung up an unleachable tank may be obtained on account of too much slime settling; if a stoppage in the work occurs a layer of slime settles on the top of the sands, which will prevent the dry leaching of that vat. These difficulties have been met by dividing the pulp and running into two more vats, as is done at the Primrose; or by adding clean water to the launder, or again, where the pulp is very muddy, by only carrying a few inches of water over the surface of the rising layer of sand, and discharging the slimy solutions by means of a rising gate discharge instead of over the top. However, where the conditions are fairly even, and the size of the vat or vats has been nicely adjusted to the work required of them, a very clean even product is obtained automatically and with no cost or attention. Examples of clean products may be found at the Robinson and at the Meyer and Charlton Mine. At the Jumpers Mine the pulp so collected is treated directly without removal, and this gives, I understand, satisfactory results; but the ordinary procedure is to remove these collected sands to the cyanide vats for treatment. In the new double treatment works erected at the May and at the Crœsus, the first solution of cyanide is directly applied to these vats, and, after draining, the pulp, wetted with cyanide solution, is transferred to the second treatment vats. It has not yet been proved whether a better result is thus obtained. I mean by better results higher profits. More cyanide is consumed on account of the complete exposure to the atmosphere, this probably being due to the oxidation of the pyrites. It is possible that double treatment may be unnecessary in this system of collecting, since the sands so collected give after transference to the leaching vats a very perfect percolation, and good results have been regularly obtained. While double treatment may very probably be found to pay where the sands are collected in a vat without the use of a distributor, as in Mr. Jennings' method, it has yet to be seen whether the method will pay upon sands collected by means of the distributor. The probable direction in which the improvements will be made will be by combining the use of the spitzluten with the distribution of the classified pulp by means of distributors into collecting vats for direct treatment. Such a plant is now at work at the central works of the Central Ore Reduction Company, and one is also being erected at the new Robinson slime plant."

Mr. Bettel writes:—"For some time past metallurgists on the Rand have been giving their adherence to the belief that, for Rand ores, close concentration—so called—of tailings is an expensive and, in a great number of cases, an unnecessary process, for the residual tailings have to undergo an additional treatment by cyanide. Mr. Henning Jennings, an early convert to the movement for abolition of Frue vanners and other concentrators, decided that it would be cheaper and in every way more satisfactory if the tailings, after separating the slimes and some fine sand which retarded filtration and draining from superfluous water, were treated with cyanide, without attempting to separate coarse material, whether sand or pyrites. Mr. Jennings' contention was, that if he obtained 50 per cent of the gold contained in the vanner concentrates, the method would be cheaper than close concentration and treatment of concentrates so obtained by cyanide and chlorination. As first introduced this process caused a loss, due to the non-extraction of a portion of the gold occurring (1) enclosed in pyrites, (2) in particles of mineralised quartz, (3) as coarser particles of gold which have escaped amalgamation, and (4) as finely divided particles of amalgam, products which require a prolonged contact with cyanide solution.

"The process, as at first proposed, is practically obsolete, all that remains of it being a rough, preliminary separation of sands from slimes, and filling the treatment tanks with tailings conveyed through hose manipulated by Kaffirs. It is now practically admitted that the time required to treat the coarse sand and pyrites is much greater than that required to treat tailings freed from such material. This the companies who used the 'direct treatment' method have shown by—1st, Introducing spitzluten for classification. 2nd, Prolonged treatment of the coarse material so produced by cyanide. 3rd, 'Double treatment' (or retreatment) of tailings."

Mr. Williams, manager of the Crown Reef Cyanide Works, writes us as follows:—"Having for a long time held the opinion that a cheaper method of classification would be advantageous, in order that the coarser particles could be subjected to cyanide according to the time required for solution of the gold, I have put this into practice with the following results, viz.: —Before starting classification our residues from cyanide works at the Crown Reef were very high, namely, from 1.6 to 1.9 dwts., averaging for three months 1.76 dwts. On making a careful examination, I found that a very large percentage of the gold was contained in the coarse concentrates and coarse sand. Analysis of these before and after treatment showed that the extraction in concentrates was 53 per cent, and in the coarse sands only 41 per cent, but, taking into consideration the cost of concentrating by the ordinary methods on a poor ore, figures proved that even this low extraction gave a better profit than could be obtained by concentration; and again, since concentration would not take out the coarse sands which carried the larger percentage of the gold, it was out of the question. I started experimenting with a spitzlutte having a supply of water at the bottom. This proved very satisfactory, and has the advantage both of small initial outlay and, being automatic, of low working costs, amounting to only £25 per month for water to classify 17,000 to 18,000 tons of ore.

"One of the most interesting points about this method is the fact that a saving of 40 per cent is effected in the cyanide consumption, as these coarse particles being removed, two strong solutions of 0.25 per cent do better work than six and seven of 0.3 per cent did before.

"With regard to filling I may state that I am perfectly satisfied with the direct method. It has the advantage of a preliminary rough concentration, by which the tailings treated are always richer in gold than the slimes, and being continually under water little or no oxidation can take place, thus a saving in cyanide is effected; but if it were possible to obtain as good a distribution of the tailings by intermediate tanks, this method has the following advantage: being automatic it saves 15d. per ton; but in many cases leaching has been impossible, whereas for the last fourteen months, treating 12,000 tons per month, I have experienced no trouble by direct filling."

Mr. A. von Gernet:—"In no case do I consider direct filling good, because the sands deposited in this way pack hard. The directly filled tank holds 25 per cent more sand than one filled by dumping dry tailings. In such a case it is very difficult to bring the particles into contact with air, and uniform leaching becomes impossible. Personally I prefer the so-called double treatment. The advantage of this method is that if the treatment is rendered imperfect in the first tank by the formation of channels, the transference of the sands to the second tank will effect a complete mixing of the material, and a thorough exposure to the air is obtained, the result being a very good extraction. The disadvantages of double treatment are:—(*First*) Great care must be taken to wash the first tank thoroughly free of cyanide, or gold will be lost during refilling with fresh pulp from the battery. (*Secondly*) The consumption of cyanide is somewhat high on account of the formation of iron salts during oxidation while transferring. Consequently if low-grade tailings are being treated, where every expense must be cut down, double treatment should not be used.

"Whether Butters and Mein's patent distributor, or filling by hand-labour with the aid of a hose, be used is not of much importance, and certainly neither method involves the exclusive use of either direct or intermediate treatment, as some people seem to imagine, but I think that mechanical distribution is preferable to manual filling, both on account of economy and of uniformity in the result."

2. *The Precipitations of the Gold Solutions.*

Mr. Bettel writes:—"With regard to the respective merits of the MacArthur-Forrest and Siemens-Halske processes for gold precipitation I must say that the zinc precipitation method seems to be the easiest worked and eminently suitable for both large and small companies. It is a disadvantage in the Siemens-Halske process that the cupellation of the lead bullion has to be carried out at a central works, and this prevents the companies being independent of a customs works. Otherwise the process is one deserving the greatest consideration as one applicable for large customs works or mines."

Mr. Williams:—"I have not had an opportunity of working with the Siemens-Halske process, so know nothing of its disadvantages (if any), but it has the following advantages over the zinc method, viz.:—

"1st, Precipitation is obtained in the very weakest solutions, thus making it possible to use much weaker solutions for leaching than is possible in the zinc method; from reliable figures I am of opinion that from 4d. to 6d. per ton can be saved in cyanide when working on the same ore.

"2nd, Acid and neutral solutions also yield up their gold, although in these cases it is impossible to precipitate by zinc.

"3rd, It is a much cleaner method, and the loss by calcination and smelting, amounting in most careful work to from ½ to 1 per cent of the gold in the zinc method, is avoided.

"4th, The royalty is just one-half."

Mr. A. von Gernet claimed, in a paper read before the Society of Chemists and Metallurgists in Johannesburg, that the most important feature of the electrical process is that very weak cyanide solutions are suitable for electrical precipitation, although difficult to treat by the chemical method. With reference to this paper, which appears to have given rise to some misunderstanding, he writes:—"The idea has become prevalent that I had said it was impossible to precipitate weak solutions by the zinc process. This is not the case, scientifically speaking, because I am well aware that with sufficiently big zinc boxes and a large amount of fresh zinc shavings, good precipitation can be obtained from weak solutions, but in practice it is difficult to maintain these conditions, and the clean-up becomes impracticable, as the gold in such cases does not form slimes, but adheres as a firm coating to the zinc, and cannot be recovered except by smelting down the zinc or dissolving in acid. Another advantage of the Siemens-Halske process is that the gold being precipitated on lead foil, a purer bullion is obtained by cupellation than is got by smelting zinc slimes."[1]

3. *The Treatment of Slimes.*

The treatment of slimes, which till recently have ranked as a waste product of gold milling, is on the eve of being practically carried out on a large scale. It is a fact already well known that the Central Ore Reduction Company have recently entered into an arrangement with the Robinson Company to treat their accumulated slimes, and that Mr. Charles Butters is erecting a large plant designed for that purpose. For the present, however, Mr. Butters is reticent on the subject, preferring to wait until he has been able to demonstrate on the commercial scale that his process is a success.

Mr. Williams informs us that he has designed a plant to treat the slimes at the Crown Reef, and furnishes the following account of his process:—

"I have designed a plan by which I am certain these

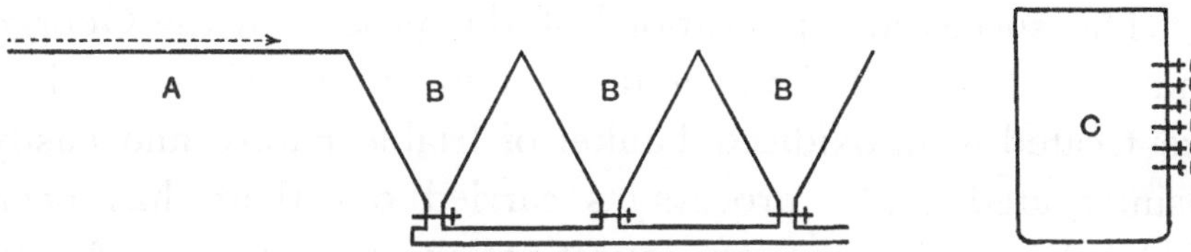

FIG. 80.—Plan for treating Slimes at the Crown Reef.

slimes can be worked at a cost of less than 4s. per ton with 80 per cent extraction.

"To the slimes water running through the launder A, sufficient lime is added to precipitate the slimes in the pits B, which are connected with a centrifugal pump, by which

[1] The gold obtained by the Central Ore Reduction Company by the Siemens-Halske process is sold directly to the Transvaal Mint, where it is easily brought to the requisite fineness for coinage purposes.

they can be pumped into the tank C; when this is full it is allowed to settle, which will take about an hour; the excess water is then drawn off by the cocks D or by a syphon. The slimes will take up about 20 per cent of the capacity of this tank; cyanide solution of .01 per cent is added in sufficient quantity to fill the tank, when it is agitated by a centrifugal pump for ten minutes, which is long enough to dissolve 95 per cent of the gold. It is allowed to settle, when about 80 per cent of the clear solution, carrying nearly 80 per cent of the gold, can be drawn off by the side cocks. I propose to use the Siemens-Halske method for precipitation, as owing to the very dilute solution of cyanide, and also the excess of calcium salts, I find it impossible to precipitate by the zinc method. Should the slimes be rich, by the addition of water and another agitation, it could be made to yield another 7 to 10 per cent. I have worked this process on a 3-ton scale with excellent results."

4. *The Direct Treatment of Rand Ore by Cyanide.*

The successful application[1] of this process at the George and May has excited much interest in mining circles. The ore treated is an oxidised banket of friable nature and easily disintegrated. The process as carried out there has been made the subject of an investigation by Messrs. A. F. Crosse and W. L. Hamilton, and these gentlemen have sent in a report to the Concentrating Committee of the Chamber of Mines, from which we take the following particulars :—

The ore is brought down from various parts of the reef, where it is mined and dumped on a coarse screen; the lumps are broken up in a No. 3 Gates crusher, and being after-

[1] By Mr. Freeman, general manager of the George and May Gold Mining Company.

wards mixed with the fine stuff from the screen, is removed in trucks to the cyanide vats, where it is treated as follows :—

1st wash .	0.06	per cent	K. Cy.	25 tons solution
2nd wash .	0.28	,,	,,	25 ,,
3rd wash .	0.10	,,	,,	40 ,,
4th wash .	0.05	,,	,,	40 ,,

then sometimes 5 to 10 tons pure water. About sixty hours are required for the whole treatment. Some lime is mixed with the ore, and on an average 0.89 lbs. of cyanide of potassium are used per ton of ore treated.

Owing to the coarse nature of the stuff it is not very easy to sample, but the ore before treatment is said to assay rather over 5¼ dwts. fine gold per ton. A fair average sample of residues from one of the vats being discharged gave 1 dwt. of fine gold per ton, and as 4 dwts. had been the average extraction, this would give a total extraction of 75 per cent.

The results for one month show an extraction on ore assaying 5 dwts. 6 grs. of 76.2 per cent, leaving a profit of about 3s. 9d. per ton. The same ore treated by amalgamation and cyaniding would have given a yield of 12s. 3d. per ton (as against 15s. 9d. by direct cyaniding), at a cost of 16s. per ton (as against 12s.), or a loss of 3s. 9d. per ton as against a profit of 3s. 9d. by direct cyaniding.

The results of amalgamation and cyaniding of tailings during six months, June to November 1894, show an actual recovery of only 55.8 per cent, from ore assaying 9 dwts. 15 grs. The loss appeared to be principally in the large proportion of rich slimes (from 40 to 50 per cent) which could not be treated. This and other reasons induced Mr. Freeman to try direct cyaniding on the poor ore available at upper level and outcrop.

With regard to *direct treatment* Mr. Butters writes us as follows :—"The direct treatment of dry-crushed gold-ores is as old as the chlorination process. It has been carried out in Australia at the Mount Morgan Mines for many years, and has been introduced in the United States, specially by the Rothwells, first in Canada by the elder Rothwell, and later by the nephew in the Golden Reward Chlorination Works of Dacotah. There are now several large direct chlorination mills running in the United States. These mills are all necessarily dry crushing mills, the crushing being generally performed by Krom rolls. The use of cyanide as a solvent for the gold in connection with Krom rolls will in some cases permit us to dispense with the roasting operation. This is especially so on the Witwatersrand, but not in the United States. As a general rule, if we omit the Witwatersrand, dry crushing has been resorted to because amalgamation was not profitable. The treatment of the dry-crushed product almost anywhere in the world would have to be followed by roasting, in order to obtain a profitable result, whether the gold was to be dissolved either by chlorine or cyanide. Chlorine is a very perfect solvent for gold, so is also cyanide, but where the gold may be a little coarse, or where the roasting may have not been absolutely perfect, chlorine has this advantage over cyanide, it dissolves the coarse gold quicker, and it has the power of finishing the roast, that is, when present in sufficient quantity, which it always must be. It decomposes the insoluble basic sulphates and the soluble sulphates, dissolves the gold, and does not leave any insoluble slimy hydrates of iron and compounds of Prussian blue as cyanide does. But given an absolutely dead roast, the solution of the gold by cyanide is just as perfect as the solution of the gold by chlorine. The recovery of the gold, however, from the chlorine solution is much more direct and positive than that obtained from cyanide solution. The use of cyanide as a solvent, generally speaking, is not nearly as prominent in other portions of the world as it is on the Witwatersrand goldfields.

"There is no doubt in my mind that the practice of the direct treatment by dry crushing of gold ores that do not amalgamate readily will be a growing practice. This is shown by the number of the mills that are being constantly erected in various portions of the United States, for instance in the Cripple Creek District of Colorado, one of the new rich gold-mining districts in the United States, which has now adopted this dry crushing and direct treatment method on a very large scale, chlorine, cyanide and bromine all being utilised by rival companies.

Taking the ordinary Witwatersrand ores as now exposed in the Main Reef Series, it is simply a question of cost of extraction as to which process will be adopted in each particular case. Assume an average ore assaying 15 dwts. to the ton as a typical case of value, either blue ore or oxidised ore, the question to be answered is how may the 60s. contained in these 15 dwts. be obtained for the least money. Your ordinary practice shows, in the first place, that this ore must be reduced in size to at least 600 mesh in order to expose the gold. This can be done at a cost, with very good work, of about 2s. 6d. per ton. The cyaniding of the tailing of the sands from this operation will be done by the Siemens' process in a plant of this size for an additional 2s. 6d. on 70 per cent of the tailings. The slimes forming the other 30 per cent will be treated at an outside cost of say 7s. 6d. per ton.

"Thus we have :—

Crushing and amalgamation . . .		2s. 6d.
Treatment of 70 per cent of the sands .	2s. 6d.	
Treatment of 30 per cent of the slimes .	7s. 6d.	
Gives an average cost of . . .		4s. 0d.
Total cost . . .		6s. 6d.

The average extraction by amalgamation . .	65 per cent.
And the average extraction by sizing and slimes treatment will be 80 per cent of 35 per cent .	28 per cent.
Total extraction . .	93 per cent.

"It is quite safe to say that the future of large, well-conducted, wet crushing stamp-mills, followed by cyanide treatment of sands and slimes, will give an average extraction of 90 per cent of the 15 dwt. ore, at a cost of about 6s. 6d. It may be expected that before the close of 1896, the majority of all the large wet crushing mills on the Witwatersrand will be obtaining over 90 per cent of their gold at a cost not exceeding 7s. per ton for metallurgical operations.

"Let us consider now the case of dry crushing upon the same ore. Mr. Stephen Krom of New York would, I think, guarantee to erect a dry crushing mill on the Witwatersrand of a capacity equal to 200 stamps that would crush the ore to 600 mesh for 2s. per ton. The treatment of this resulting pulp is not so simple a problem as might at first sight appear. It contains coarse gold which is ordinarily sized out by amalgamation; it contains the pyrites, the coarse particles of which are sized out by spitzluten, and the coarse and fine sands and slimes resulting from crushing. Theoretically this stuff should be sized in order to reduce the time and expense of treatment. It can be sized by air: if this be done, you have your slime resulting product, which is unleachable, and must be treated as wet slime is treated. If sizing has to be resorted to after dry crushing in order to produce as high an extraction (the same extraction for the same cost) as with wet crushing, I do not quite see the advantage in dry crushing. This is the sticking-point. If this pulp can be leached at once without sizing, the treatment will necessarily be much longer in order to get the same extraction on account of the coarse gold and pyrites, therefore more expensive; and I do not think it possible to obtain the same extraction on a large scale on unsized pulp that can be got on sized pulp. If, however, you resort to the usual time of treatment upon this product directly, and then wash this treated pulp over an amalgamated plate and through spitzluten to get the coarse gold and a second product for treatment, you are sure to have unavoidable loss of soluble gold on account of the action of the dilute cyanide solution left in the product after treatment on the still undissolved coarse gold and on the incompletely treated pyrites. Thus subsequent amalgamation and sizing of the treated pulp would, I think, be impracticable on account of this loss of gold; and therefore, unless a good extraction could be obtained by a direct solution of the gold in this crushed pulp without other operations, I do not see how the dry crushing of the average ore could compete with wet crushing. If the same extraction upon this pulp is to be obtained merely by one operation, it will cost at least 4s. per ton on account of the increased time necessary for solution. This would bring the total cost of dry crushing and direct treatment up to 6s. per ton as against 7s. by wet crushing.

"The advantages to be gained are so slight, and the difficulties to be overcome so great, that I think it will be many years before it is attempted on a large scale on these fields.

"The above remarks simply refer to the average ore of the Witwatersrand which we are now working. No doubt there are many special cases in which the ore does not need to be finely crushed in order to expose the gold; in such cases dry crushing and direct treatment may be more profitable than wet crushing."

Mr. Bettel writes:—"There are, undoubtedly, certain surface ores on the Rand containing gold in an extremely fine state of division which are amenable to 'direct' treatment by cyanide. In such cases, where the cement of the conglomerate is so friable and porous that a comparatively coarse crushing through Blake-Marsden stone-breakers or Gates crushers is sufficient to pulverise the gold-bearing material, leaving the comparatively barren pebbles in large fragments, I consider the process an admirable one.

"Where, however, the hardness and non-porosity of the cement is such that the whole of the material has to be crushed to a comparatively even size, I fail to see the advantage of the proposed method over the present system, for the following reasons :—

1. The treatment of slimes produced in the ordinary wet process presents very few difficulties, and may form part of a continuous process.
2. By wet crushing double the amount of ore may be crushed per horse-power than is possible by dry crushing.
3. Coarse gold is recovered by amalgamation easily and cheaply in the ordinary method of wet crushing.
4. Classification and 'double treatment' are possible in one series of operations in wet crushing.
5. If desired, after treatment of coarse sands and pyrites (produced by spitzluten) by cyanide, the residues may be concentrated (say by Gowan's concentrator), and a further portion of gold extracted from the concentrates by retreatment with cyanide at a comparatively small cost.
6. The process of dry crushing and direct cyaniding without further treatment is inapplicable to the majority of Rand ores on the score of low extraction.
7. The alleged saving of cyanide by dry crushing is mythical, as under the most favourable conditions yet quoted about 1 lb. of cyanide per ton is consumed in the direct treatment, while with the ordinary MacArthur-Forrest process at the Robinson Mine only 0.41 lb. of cyanide is consumed, and by using extremely dilute solutions, as in the Siemens-Halske process, the consumption is only about $\frac{1}{4}$ lb. per ton.
8. In fine crushing (dry), the ore, if wet, has to be deprived of its moisture, at a cost (apart from handling) proportional to the moisture to be removed.
9. In fine (dry) crushing of deep level ores, the amount of

finely divided ore present prevents by 'packing' a satisfactory leaching of the pulp.

10. In the treatment of hard banket, owing to the excessive wear of grinding surfaces, in fine dry crushing an electro-magnetic separator will have to be used to remove excess of finely divided iron. This machine may at the same time remove gold present in magnetic particles of ore (magnetic pyrites, magnetite, etc.)."

Mr. Williams writes:—"All experiments tried by myself on ore from the Main Reef Series by direct cyaniding (without previous amalgamation) have been a failure. An experiment on a 14 dwts. pyritic ore from the Crown Reef showed an extraction of only 57 per cent after treatment equal to 14 days. Of the 6 dwts. left in the residue, 5¼ dwts. was free gold which could be recovered by amalgamation, but I would point out that amalgamation after cyaniding is not practicable. In ores where the gold is in a *very fine state* and the cement porous, there may, however, be scope for this process."

And Mr. A. von Gernet:—"I have only had experience of direct treatment of ore from two places, one of which was on the Kimberley Series, and the other on the Black Reef. In both cases the ore was oxidised and very friable. By screening through a half-inch mesh 50 per cent of the matrix was obtained, the rejected pebbles being practically valueless. In the case of the Black Reef sample a 75 per cent extraction was obtained at once, but the sample from the Kimberley Series gave an extraction of less than 10 per cent after a 7 days' treatment both with strong and weak solutions. This remarkable result clearly shows that the character of the ore has a considerable influence on the result obtained. In some cases, no doubt, the process will be useful, but in my opinion it will never attain to the importance that has been claimed for it."

PARTICULARS OF THE CYANIDE PROCESS AS PRACTISED AT SOME OF THE PRINCIPAL CYANIDE WORKS ON THE RAND.

1. *Robinson*[1] *Cyanide Works.*

The cyanide plant at the Robinson Mine is capable of treating 6500 tons of tailings per month. The pulp leaving the mill is elevated by a bucket tailings wheel, and the sands are collected by means of Butters and Mein's Patent Distributor in intermediate vats, whence they are transferred to the treatment vats. The process in use is the MacArthur-Forrest, the gold solution being precipitated by means of zinc shavings. The extraction is 70 per cent. The residues after treatment assay 2 dwts. The consumption of cyanide is 0.5 lbs. per ton treated. The working costs per ton exclusive of royalty amount to 3s. per ton, the items being as follows :—

Average Working Costs for a Period of Six Months.

	S.	D.	
Wages (including natives' food)	0	8.3	per ton.
Stores (assay, smelting, materials, etc.)	0	2.9	,,
Fuel and water	0	2.7	,,
Cyanide	1	0.9	,,
Zinc	0	0.6	,,
Filling and discharging tanks	0	9.4	,,
	3	0.8	,,

The bullion obtained has an average fineness of 778, and value of 64s. 6d. per ounce.

2. *Crown Reef Cyanide Works.*

Mr. Williams, manager of these works, informs us that three products are being treated by cyanide :—

"First, 5 to 6 per cent of the coarsest particles of con-

[1] Supplied by Mr. G. A. Darling, manager of the Robinson Cyanide Works.

centrates and sands are first taken out and treated. This product assays 1 oz. per ton, and contains about 33 per cent of pyrites; an extraction of 90 to 92 per cent is obtained at a cost of 6.2 shillings per ton, exclusive of royalty.

"Second, 7 to 8 per cent of the next coarsest material is extracted. This product assays 9 to 10.5 dwts. An extraction of 92 to 95 per cent is obtained in seven to eight days at a cost of 3.5 shillings per ton, exclusive of royalty.

"Third, The balance of the sands with some slimes are first treated in the direct tank with one solution of 0.25 per cent potassium cyanide; after they have been drained dry, they are transferred to another tank where leaching is completed. We are now getting an extraction on these sands of 80 to 85 per cent at a cost of just under 3 shillings per ton, exclusive of royalty.

"The balance, amounting to about 26 per cent of the ore milled, is slimes; this, however, does not carry more than 7 to 8 per cent of the original gold in the ore, or from 15 to 18 per cent of the gold leaving the mill. These slimes have up to the present been stored for future treatment. The average assay value of the residues after treatment was (according to the last annual report) 1.367 dwts. for the sands, and 1.730 for the concentrates."

3. *City and Suburban Cyanide Works.*[1]

The tailings from the 80-stamp mill are elevated by a tailings wheel and passed through hydraulic separators. Three classes are obtained, which are combined for one treatment. The assay value of the combined sands is about 15 dwts. per ton; that of the slimes about 4 dwts.

The capacity of the cyanide plant is 15,000 tons per month. The tanks are filled direct. The treatment occupies ten days. The extraction is about 62 per cent. The residues

[1] 1st June 1895.

after treatment average 1.9 dwts. per ton. The precipitation is by MacArthur-Forrest process. The consumption of cyanide is .649 lbs. per ton treated; that of zinc .228 lbs. per ton, that of caustic soda .006 lbs. per ton. The working costs average 4s. 2.7d. per ton. The value of the bullion obtained is £3 : 5 : 7 per ounce.

4. *New Primrose Cyanide Works.*[1]

Mr. W. Bettel, consulting chemist to the New Primrose Mine, communicates the following notes on the cyanide works of this mine:—

"At the New Primrose Company's Mine, prior to January of this year, the product from the 160 stamps, without concentration or classification, was treated in the plant previously used for the sands from the 100 stamps, with the result that only 55 per cent of the assay value was extracted from the tailings. It was therefore necessary to introduce a quicker method of dissolving the gold. I therefore introduced hydraulic classifiers (spitzluten), making all necessary alterations in pipes, etc., to facilitate rapid transference of solutions, using vacuum pumps for facilitating drainage and replacing the wheel valves on the intermediate tanks (water drainage system) by Bell's asbestos-lined plug valves. A vacuum of 20 inches mercury was thus readily obtainable, but 15 inches proved more convenient in practice, as with the higher vacuum the filter cloths were forced by atmospheric pressure, etc., to the bottoms of the tanks, and impeded filtration. In outline the method adopted was—Classification: about 10 per cent of 15 dwts., sands and pyrites being collected. This product after draining was treated for three or four days with 0.3 per cent of cyanide solution, then drained and discharged

[1] The capacity of the New Primrose plant is 17,000 tons per month. The process used is the MacArthur-Forrest. The consumption of cyanide per ton is .98 lbs.; of zinc, .22 lbs., and of caustic soda, .22. The working expenses average 4s. 1d. per ton. The value of the bullion recovered is 65s. per ounce.

to final treatment vats, where it underwent from 2½ to 3 days leaching and lixiviation, followed by water washes. The extraction was 85 to 93 per cent, the consumption of cyanide about 1½ lbs. per ton.

"The pulp leaving the classifiers was received in distributing tank, and conveyed thence to intermediate tanks fitted with circular baffle plates[1] at the top near the rim-discharge. The tanks after draining and being 'vacuumed' had their contents levelled at the surface, then all cocks being closed, a solution of 0.04 per cent strength was run on, drained, another and stronger solution put on and allowed to stand 12 hours; the sands were then drained under about 5 to 7 inches vacuum, discharged to ordinary leaching tanks, and treated with a 0.2 per cent solution, which after standing for 12 hours was allowed to percolate, 0.08 per cent solution, and 0.04 solution being used successively, finishing with a water wash rendered alkaline with milk of lime. The solution from the pumps passed through four filters filled with coir fibre before passing to the zinc boxes, the slimes being collected by a filter-press.

"The results yielded by the modifications known generally as 'double treatment' are very satisfactory—75 to 85 per cent being obtained, an average of 78 per cent produced as against 55 per cent by the old method, showing the remarkable difference obtained by the new treatment. It may here be remarked that the final residues show that from 15 to 18 grains of gold are enclosed in sands and pyrites, in such a manner that only regrinding will liberate the gold to render it accessible to solvents. Further, the classification admits of the use of 700-mesh screen against a 900 previously used, resulting in the increased output by ¾ ton per stamp per 24 hours, or 3600 per 30 days without extra cost for labour or fuel in the mill."

[1] The baffle plate is used to prevent the loss of coarse and medium sands with the slimes.

5. *New Heriot.*

The capacity of the plant is 5000 tons per month.

The tailings elevated by a bucket-lift are separated into three classes by means of spitzkasten :—

1.	(Concentrates)	assaying	32.93 dwts.
2.	(Sands)	"	5 03 "
3.	(Slimes)	"	3.35 "

The first two classes are treated by cyanide, the slimes being stored for future treatment.

The concentrates are submitted to 13 days' treatment, an extraction of 87.51 per cent being obtained.

The sands receive 5 days' treatment, and the extraction is 69.5 per cent.

The residues from the sands assay 1.54 dwts. The bullion recovered is of a value of 60s. 9d. per ounce.

The consumption of cyanide per month amounts to .84 lbs. per ton treated, that of zinc to .30 lbs. per ton, and that of caustic soda to .20 lbs. per ton.

The working expenses average 4s. 2d. .25 per ton treated.

6. *Worcester Cyanide Works.*[1]

The battery pulp leaving the plates and Frue vanners passes through spitzluten and spitzkasten, the former to take out the coarse sand and pyrites, and the latter to separate the fine sands from slimes.

There are formed four different products :—

1.	Coarse sand and pyrites	assaying	15 dwts.
2.	The bulk of the sands	"	6 "
3.	Fine sand	"	4½ "
4.	(Unleachable) slimes	"	4½ "

The first product amounts to 15 per cent of the pulp, and is treated for nine days with 0.08 per cent cyanide, giving 1½ to 2 dwts. residues.

[1] Communicated by Mr. A. von Gernet.

The second product amounts to 50 per cent of the pulp, and leaves, after five days' treatment, 1 to 1.25 dwts. residues.

The third product amounts to 10 per cent of the pulp, leaving 1 dwt. residue.

The fourth amounts to 25 per cent, and is not treated at all.

The Siemens and Halske electric precipitation process is in use. The consumption of cyanide at the works amounts to ¼ lb. per ton treated. 3000 tons are treated per month at a cost of 3s. per ton.

Sufficient has been said about this plant in previous publications.

7. *Metropolitan Cyanide Works.*[1]

The pulp leaving the plates of the 60-stamp battery is lifted by a tailings wheel into a series of spitzluten and spitzkasten. The pulp contains 3.25 dwts. of gold.

It is sized into three products :—

25	per cent of it assays	5	dwts.	(coarse sand).
75	,,	,, 3	,,	(bulk of the sand).
30	,,	,, 1.25	,,	(slimes).

The tailings are hauled by electric rope-haulage to the leaching vats, and treated 5 days with a 0.08 per cent solution; the residues run 1 dwt. to 12 grains. The working cost is about 2s. 10d.; electrical precipitation is adopted.

8. *May Consolidated Cyanide Works.*[1]

The pulp of the 80-stamp mill is raised by a double cylinder plunger pump, 55 feet high, into a launder, and flows for about 100 yards to the cyanide works. There the tailings are deposited in the tanks by means of Butters and Mein distributors. The tanks are carried on wooden

[1] Communicated by Mr. A. von Gernet.

trestles 20 feet high, containing 200 tons each. There are four upper tanks and four lower tanks, in order to give the sands a double treatment. The tailings assay about 4.5 dwts., the residues 21 grains on the average.

The consumption of cyanide is .3 lbs. per ton of tailings. Electrical precipitation is in use. The working expenses are about 2s. 6d. per ton treated.

9. *New Crœsus Cyanide Works.*[1]

There are five collecting vats, five leaching vats, and two vats for concentrates. A 45-feet tailings wheel lifts the pulp; after passing through two spitzluten to separate the concentrates, the tailings are carried by Butters distributors into the tanks. Double treatment and electrical precipitation are in use. As the works have only been running for a month, no data with regard to results can be furnished.

[1] Communicated by Mr. A. von Gernet.

CHAPTER X

THE ECONOMICS OF THE MINES

Material and Supplies

On the discovery of the Witwatersrand goldfields there was no railway in the Transvaal, the nearest terminus being Kimberley, on the northern border of the Cape Colony, some 300 miles distant. Since the country itself actually produced nothing, all material, mining and other supplies had to be brought by ox or mule waggon from that point at very heavy cost. In 1889 the average cost of transporting goods from the coast to Johannesburg was £30 a ton, of which £8 was railway freight, and the balance consisting of charges for waggon transport, including loading, "off-loading," and agency. The rapid increase of the population during 1888 and 1889, together with the very inadequate facilities for transport, gave rise in the latter part of the winter of 1889 to a rather ominous scarcity of provisions. Although no serious consequences resulted, owing to the immediate steps taken by the Chamber of Mines and the Governments of the South African Republic, Natal, and the Cape Colony, the evil of a recurrence of the same or worse state of things was realised. Adequate measures were taken to ensure the forthcoming of ample stores in case of emergency, and pressure was brought to bear on the Government in the matter of railway extensions, to which but little encouragement had been given by the Boers up to that time. In the following year, 1890, the extensions of the Colonial line in the Free State and the internal railway from Pretoria to the Vaal River were commenced and rapidly pushed on, and in the latter end of 1892 connection was made between the coast and Johannesburg, when the highly-appreciated advantages of regularity and cheapening supplies were immediately felt.

At the present time the freight from the coast by Cape Government and Netherlands Railways for mining machinery, material, etc., are as follows :—

3rd Class Goods, including Machinery and Mining Material.	Per Ton.	Per 100 lbs.
	£ s. d.	s. d.
Cape Town to Johannesburg	9 3 4	9 2
Port Elizabeth to ,, ...	7 3 4	7 2
Grahamstown to ,, ...	6 16 8	6 10
East London to ,, ...	6 15 0	6 9

Rough goods are charged about £2 more per ton.

The shipping freights from England by the Castle and Union Lines intermediate steamers are as follows :—

	Machinery in pieces under 40 cwt.
To Cape Town and Port Elizabeth . .	32s. 6d. per ton.
To East London and Natal	40s. ,,

Special arrangements are made for cases weighing over 40 cwt. The charges by mail steamers are about 1s. 3d. more all round.

Machinery freights from New York vary from 30s. to 150s. per ton. The bulk of the goods imported are landed at a cost of from 40s. to 50s. per ton, increased rates being charged for cases weighing over 30 cwt.

The customs tariff for imported goods has always been high ; and although, through persistent protest on the part of the foreign element in the State, some of the more burdensome taxes have been reduced, the weight borne by the industry is still unduly heavy, as the following list will show:—

	Ad valorem.[1]	Additional special duty.
Sulphuric acid .	7½ per cent.	1d. per lb.
Candles . .	,,	None.
Cement . .	,,	3s. per 100 lb.
Coke . .	,,	,, ,,
Dynamite . .	,	Not exceeding 9d. per lb.
Corrugated iron .	,,	,, ,,
Chains . . .	,,	,, ,,
Bar iron . .	,,	,, ,,
Rod ,, . .	,,	,, ,,
Sheet iron . .	,,	,, ,,
Oil . .	,,	,, ,,
Sheet lead . .	,,	3d. per lb.
Machinery and component parts thereof[2]	1½ per cent.	,, ,,
Pitchpine and other timber used in construction works .	,,	,, ,, ,,
Quicksilver . .	,,	,,
Cyanide . .	,,	,,

In the following table will be found the present local prices of the chief articles of consumption and material employed in mining. The third column contains estimates of the total quantities consumed at sixty-seven working mines in the district in 1894,[3] and the fourth column the average prices paid during that year:—

[1] The *ad valorem* duty is reckoned on the English price *plus* 20 per cent to cover freight and other charges.

[2] Certain other mining material is included under this head.

[3] These estimates are taken from the *Chamber of Mines Report* for 1894, he returns having been sent in by 67 companies alluded to.

[TABLE

Article.	Present Prices, May 1895.	Quantities consumed in 1894.	Average Price in 1894.
Cement*	46s.	12,033 casks	53s. 4d.
Deals, 3″ by 9″	8d.	2,528,193 feet	9d.
Galvanised iron	6½d. per sq. ft.	...	...
Candles*	13s. 3d.	93,064 boxes	15s.
Dynamite*	85s.	47,938 cases	88s.
Blasting gelatine	107s. 6d.	29,361 cases	105s. 6d.
Detonators*	2s. 9d. per box	...	...
Fuse*	5½d. to 8d. per coil	...	...
Shovels	3s. to 3s. 4d. each	...	...
Hammers	5½d. per lb.	...	...
Drill steel	5d. to 7½d.	1,562,143 lbs.	6⅞d.
Rails	270s.	1433 tons	324s.
Bar iron	2d. to 2⅛d.	...	...
Quicksilver*	190s. to 195s.	649 bottles	160s.
Mealies and meal*	16s. 6d. per muid	150,067 muids	14s. 6d.

* Cement, a cask contains about 380 lbs. net.
Candles, a box contains 25 lbs. net.
Dynamite, a case contains 50 lbs.
Detonators, a box contains 100.
Fuse, a coil is 24 feet.
Quicksilver, a bottle contains about 70 lbs. net.
Mealies, a muid sack contains 200 lbs.
Mealie Meal, a muid sack contains 180 lbs. about.

Timber.—The bulk of the heavy timber imported for the mines is pitchpine from the United States. The structures in which pine is chiefly employed are headgears, battery frames, rock-houses, and shaft timbering, while for pump rods it is invariably used. The price is from 5s. to 6s. per cubic foot, according to scantling and length. Deal, which was formerly employed to a much greater extent than now for purposes for which pitchpine is better adapted, is still imported in great quantities for light construction work, buildings, etc. The cost of 3 by 9 inch deals is from 7¾d. to 8¼d. per running foot. Other woods imported for heavy and permanent timber work are Karri, a variety of Eucalyptus from Australia, and Oregon pine. Karri is a very valuable wood, extremely durable and of great strength, suitable for mortar blocks, mud sills, shaft timbering, stamp-guides, and a variety of other uses. It costs in Johannesburg about 5s. 6d. per cubic foot. Oregon pine costs from 4s. to 4s. 6d. per cubic foot.

Native timber[1] in great variety is used for mining poles and props, sleepers, fencing, etc. The price, varying with the bole, length, condition, and class, is from 9d. to 1s. 3d. per foot for poles of from 6 inches to 9 inches in diameter. Two or three kinds of native wood—as iron-wood, sneeze-wood, and blacklead wood—are from their hardness specially adapted for guide blocks of batteries.

The scarcity and general inferiority of native timber have led to the planting of immense numbers of trees in the neighbourhood of the mines, which will doubtless in the course of from five to ten years yield valuable timber at very much lower costs than these ruling at present. The plantations, which are numerous and cover large areas, consist almost entirely of species of Eucalyptus, to which the soil and climate appear to be well suited. Firs and allied species of the slower-growing classes of timber have also been planted to some extent.

Dynamite.—The interference of the State in the matter of the importation and manufacture of explosives has from the very early days been one of the great bugbears of the mining industry. The first difficulties related to the concession for the monopoly of manufacture in the State and the prohibition of imported explosives. In 1889 complaints were made to the Government, through the Chamber of Mines, against the Transvaal dynamite as compared with that which could be imported, such as Nobel's. The principal grievances were that it was both dangerous and wasteful in point of time, owing to the excessive amount of fumes produced, and that it was expensive. The danger attending its use was also due to partial explosions and

[1] The following are the names of some of the more important species of South African timber:—Camel-thorn, Sweet-thorn, Knoppys-thorn, Karll, Beuken-hout, Olyven-hout, Yellow-wood, Hartekol, Tambouti, Sneeze-wood, Stinkwood, Syringa. Much information as regards the timber used on the Rand can be obtained from a paper on the subject read by Mr. A. S. Boucher before the South African Association of Engineers and Architects, Johannesburg.

missed holes. The price charged at that time was £5 : 2 : 6 per case,[1] whereas Nobel's dynamite could have been delivered at £4 : 10s., inclusive of an import duty of 7s. 8d. per case. In 1890 a commission was appointed by the Government to inquire into the validity of the complaints urged, but no advantage resulted to the industry. In 1892, however, a determined attack was made on the concessionaires, on the grounds that the material imported by them for manufacture of dynamite was in itself explosive, was in fact *guhr impregne* containing nitro-glycerine, and that no *bona-fide* manufacture of explosives had ever been carried on by them. This point was gained, and in August 1892 the concession was cancelled by the Volksraad. As the result of the competition which ensued, the price of dynamite would have fallen as low as 85s. per case, had not the Government immediately imposed a special tax of 10s., making the price 95s. per case. Then came a State monopoly, and the appointment of a Government agent for a term of fifteen years—practically another concession. Among the provisions of the monopoly is the stipulation that within two and a half years a factory shall be erected within the State for the supply of dynamite. Meanwhile there is a special duty on explosives of 8½d. per lb., and 7½d. per cent *ad valorem*, the total tax amounting to 37s. 6d. per case. The present price of dynamite is 85s. per case, and that of blasting gelatine 107s. 6d. per case.

That an industry which has raised the State from a condition of indigence to one of affluent prosperity should be handicapped by the heavy taxation of an absolute necessity is clearly not only a piece of rank injustice to the industry, but also bad policy on the part of the Government. Hitherto no effort has been spared by the representatives of the industry in opposing and protesting against the action of the Government, and although but little advantage has been

[1] A case contains 50 lbs. of dynamite cartridges.

gained so far, it is confidently hoped that ultimate success will be achieved.

Coal.—Since the discovery of the banket formation a number of coal deposits have been opened up in the southern Transvaal that previously were either unknown or attracted little attention. It is hardly necessary at this date to emphasise the importance to the Witwatersrand goldfields of the existence of coal in their immediate neighbourhood.

The first coal used on the Rand came from the Orange Free State, but very shortly after the birth of the industry large seams were discovered in close proximity to and even overlying the conglomerates. The first local coal discovered was at Boksburg, some twelve miles east of Johannesburg, where a seam was intersected accidentally in the course of prospecting for the eastern extension of the Main Reef. The Boksburg Collieries were followed by those at Brakpan and at the Springs farther east, and later by the Cassel Colliery on Daggafontein in the same district. Since the Boksburg discovery in 1887 the coal production has increased *pari passu* with the gold production of the Rand, the Brakpan, Cassel, and Boksburg Collieries constituting the main sources of the fuel consumed in carrying on gold mining operations.

The local fuel, although much inferior in quality to most English coals, makes very serviceable boiler fuel. Like most South African coal it contains a high percentage of ash and a good deal of sulphur, with a low percentage of bituminous matter. The ash is seldom under 8 or 9 per cent, the average being nearer 15 per cent. The calorific power varies from 8 to 10. The consumption in efficient steam plants is 5 or 6 lbs. per horse power per hour, anything under 4 lbs. being very exceptional. The evaporative duty obtained varies from 4 to 6 lbs. per lb. of fuel. Being naturally slow burning, local coals require a strong draught and constant attention in stoking. The Boksburg coal is not equal in quality to that mined by the Coal Trust and Cassel Companies, but being nearer to the goldfields it is in some cases found more economical owing to the high railway rates.

The production of the principal collieries supplying the Witwatersrand in 1894 was as follows :[1]—

Transvaal Gold Trust Company (Brakpan) . .	284,432
Cassel Colliery Company	135,169
Boksburg (4 companies)	81,821
	501,422

Of this total probably about 450,000 tons were consumed on the gold mines. The coal produced in the Middelburg district, which is perhaps the best in the Transvaal, and that of the Vaal River district, are practically excluded from the Johannesburg market by prohibitive railway rates.[2] The charge made by the Netherlands Railway Company is excessive, viz. 3d. per ton per mile. The average price paid for coal in 1894 along the Rand was 18s. per ton. Of this amount, as pointed out by the chairman at the last annual meeting of the Chamber of Mines, 9s. represents transport, including waggons, loading and off-loading for an average distance of probably less than 15 miles.

1 *Chamber of Mines Report*, 1894.

2 The following statement shows the yield from, and the work done in the coal-mines of the South African Republic during the year 1894, according to returns received by the State Mining Engineer. The collieries mentioned above as supplying the Witwatersrand goldfields are here all included under the head of Boksburg :—

Mining Districts.	Number of Mines.	Tons.				Value in £ sterling.				Average number of Workmen.	
		Ex-tracted.	Sold.			Rough Coal.	Nut Coal.	Slack Coal.	Total.	White.	Col-oured.
			Rough Coal.	Nut Coal.	Slack Coal.						
Boksburg	13	614,241	422,913	166,410	2,946	206,236	47,708	1,227	255,171	124	2,005
Vereeniging . . .	1	177,594	164,051	14,590	..	89,672	3,594	..	93,266	61	702
Middelburg . . .	11	11,040	10,492	20	33	5,113	10	2	5,125	8	96
Schoonspruit . .	1	5,713	5,543	48	18	3,966	25	9	4,000	4	42
Standerton . .	4	1,850	1,850	..	..	1,040	..	..	1,040	3	16
Pretoria . . .	1	2,444	2,444	..	..	1,092	..	..	1,092	2	12
Totals . . .	31	812,882	607,293	181,068	2,997	307,119	51,337	1,238	359,694	202	2,873

Smithy coal, which is selected or brought from better seams at a greater distance than steam coal, costs on an average 42s. per ton.

Labour.

White Employees.—The class of white labour available on the field is improving yearly, as might be expected when the inducements offered are considered. The climate is excellent, the fields are of proved permanence, wages are good, and there is a further inducement in the conditions of work, the amount of manual labour falling to the lot of the white hands being reduced to a minimum, owing to the employment of natives.

The wages paid for various classes of skilled labour are as follows :—

Miners . . .	14s. to 15s. a shift.
Rock Drillmen .	£1 a shift or contract.
Mill hands (senior) .	£25 to £30 a month.
,, ,, (junior)	£20 to £25 a month.
Engine-drivers .	14s. to 18s. a shift.
Fitters . . .	15s. to 20s. ,,
Carpenters . .	,, ,, ,,
Blacksmiths . .	,, ,, ,,

General hands, with no particular trade, or learners may earn from £10 to £20 a month. The number of whites employed on the Witwatersrand Goldfields during the first quarter of the present year was, according to the State Mining Engineer's Returns, 5783.

The following is a statement from the State Mining Engineer's Report for the year 1894, showing some vocations and trades of those employed at the Witwatersrand Mines, with their average working hours and wages paid :—

Vocation or Trade.	Witwatersrand.			Heidelberg.			Schoonspruit. (Klerksdorf and Venterskron.		
	Number of Persons employed.	Wages. Average per month per Man.	Wages. Total in December 1894.	Number of Persons employed.	Wages. Average per Month per Man.	Wages. Total in December 1894.	Number of Persons employed.	Wages. Average per Month per Man.	Wages. Total in December 1894.
		£ s.	£		£ s.	£		£ s.	£
Shift Bosses	110	31 16	3,495	2	32 10	65	7	28 14	201
Miners and Trammers	1175	20 11	24,127	20	20 14	414	44	26 13	1173
Machine Drillmen	578	25 16	14,923	...	...	...	5	25 4	126
Sinkers and Pitmen	138	16 17	2,322	...	...	...	1	22 0	22
Engine-drivers	538	24 4	13,028	11	21 2	232	20	20 18	418
Pump Men	73	23 4	1,693	...	...	...	4	22 10	90
Stokers	76	17 2	1,301	...	...	...	...	...	...
Carpenters	485	24 11	11,914	6	22 16	137	11	24 11	270
Blacksmiths	440	26 4	11,532	5	24 16	124	13	24 15	322
Mechanics and Fitters	448	25 4	11,286	5	28 4	141	18	25 1	451
Painters	24	20 5	386	...	...	...	...	...	...
Masons	106	25 8	2,691	3	22 6	67	2	26 0	52
Labourers	104	18 2	1,882	...	...	...	2	20 0	40
Mine and Store Clerks	177	21 6	3,771	2	25 0	50	8	22 18	183
Amalgamators	276	23 14	6,542	3	22 0	66	18	26 17	483
Cyaniders	191	24 15	4,724	3	22 0	66	15	26 15	401
Concentrators	16	21 0	336	...	...	...	..	...	...
Vannermen	41	20 1	823	...	...	...	...	...	...
Furnace Men	23	21 10	563	...	...	...	1	26 0	26
Miscellaneous	316	16 15	5,297	9	30 4	272	16	20 15	332
Totals	5335	23 0	122,736	69	23 14	1634	185	24 16	4590

Native Labour.—All purely manual and certain kinds of skilled work are performed by natives. The classes of work for which they are mainly engaged are: in the mines—hand-drilling, shovelling, filling, tramming, also assisting machine drillmen, track layers, timbermen, etc.; on the surface—landing, dumping and filling trucks, tramming, ore-sorting, stoking and assisting enginemen, carrying coal, lumber, etc., pick and shovel work, assisting millmen, filling and emptying tailing vats, and generally all work carried on under strict supervision. In some cases natives, and more often half-castes and Indian coolies, have learned a trade and work as mechanics, smiths, engine-drivers, etc.

The South African natives are on the whole a very docile and a fairly intelligent class. They make very good strikers in the mines, and when they have mastered routine, in whatever line they are employed they are capable of doing a good day's work if properly looked after. Labour of this class on the Rand cannot be considered expensive when compared with that obtainable in most mining districts; there is, moreover, the probability that wages paid to natives will eventually be reduced, through the unremitting efforts of the Chamber of Mines which, since its inception, has directed much energy to the question of the reduction of native wages and the increase of the supply of native labour.

The history of the labour question is briefly as follows:—

In the early days the rate of wages grew rapidly owing to the inadequacy of supplies, until in 1890 the average pay of natives was as high as 63s. a month, exclusive of keep. In that year an endeavour was made to secure the concerted action of various mining companies in reducing the wages, and a scale of monthly payments to natives was agreed to by 66 companies, with the result that in the course of three months the average wage was reduced to 41s. 6d. a month. Little effort, however, was made at that time to increase the supply, and the reduction was to a great extent temporary, owing to the failure of many of the companies to maintain the scale agreed upon as soon as natives became scarce. Although wages did not at once return to the former high figures, the demand has since that time steadily increased to a greater extent than the supply, and wages have consequently again become as high as 60s. Towards the close of 1893 the Chamber, adopting a policy more likely to prove successful, instituted a Native Labour Department, with the objects (1) of assuring an adequate and regular supply of native labour, by opening up the sources of supply which exist within the Republic, and if this were found insufficient, by arranging for the introduction of natives from the east coast, and (2) of taking steps for the gradual reduction of native wages to a reasonable level.

This policy is still being pursued, and although the result has not yet been felt, it is probable that eventual success will be attained, by increasing the facilities for natives travelling to and fro, and by inducing them to make contracts binding them to stated periods of service at fixed rates. To meet the wishes of the Government the first endeavours were directed towards increasing the supply from the northern districts of the Republic.[1] A depot was established in 1894 at Pietpotgieter's Rust to tap the Zoutpansberg and Waterberg districts, which were thickly populated with natives; and at Zandfontein, a place about half-way to Pretoria, a shelter was provided, while at Pretoria another depot was placed at the disposal of the Chamber, to which the natives go prior to their departure for the Rand. Recently four buildings at suitable distances along the road have been purchased, so that during the cold weather the natives may find shelter at the close of each day's march. Pietpotgieter's Rust is the rallying point for natives desiring to come to the fields from the northern districts under the ægis of the Chamber. The white man in charge obtains for them a travelling pass and vaccination certificate; they are then fed and housed until a sufficient number have arrived to make up a travelling party, when they proceed under escort to their destination, getting food and shelter at the various stations on the way. From Pretoria, where a white man is also in charge, they are taken to the Rand by rail. A depot is on the point of completion at Johannesburg which will serve as a distributing centre. "Boys" who have worked for a company, if well treated, will prefer to return to it rather than to serve under a new employer, and if taken anywhere else against their will, will run away at the first opportunity and seek their old masters. On arrival at the Rand they will,

[1] We are indebted to Mr. A. R. Goldring, secretary of the Chamber of Mines, for his courtesy in supplying much of the information in connection with this subject.

therefore, be asked whether they wish to go to a particular company, and if a preference be expressed they will be sent there; in other cases they will be drafted off to one or other of the companies requiring labour. The system inaugurated by the Chamber is calculated to encourage the natives to come to the fields in increasing numbers, as under it not only are they guarded against the physical hardships of hunger, fatigue, and cold, which previously they had to endure, but are protected against interference and annoyance on their way down and robbery on their way back; while on the other hand they arrive thoroughly fit for work, instead of having to rest and be fed up for several days before starting on their duties.

A fair amount of success had been attained when the recent expedition against Magoeba, a native chief in the Wood Bush, was undertaken; this has temporarily disorganised the supply, as from a fear of the possible extension of the area of hostilities the natives are not allowed by their chiefs to leave their homes.

Arrangements are now in progress for establishing depots at Komati Poort and Nelspruit on the eastern route, to organise supply from the Eastern Transvaal and the Portuguese territories.

In due course the system will be extended, with the object of developing and organising the supply from the various other native centres, so that while the steadily increasing demand of the mining companies may be met, the rate of wages may be gradually reduced. Although, as has been mentioned, the cost of labour is not high as compared with other countries, the rate of wages on the Rand is considerably higher than at De Kaap and Lydenberg. This is due to the demand hardly ever falling below and often exceeding the supply. By organisation it is expected that the balance will be recovered. Were the current wages 40s. per month natives would willingly engage at that rate; but it is found impossible to get them to work at less than the average current wages, even though under contract to do so. A reduction to be effective must be general. This is the object sought to be obtained, and which, when achieved, will mean a saving of over half a million a year in native wages.

The efficiency of native labour is at the present time much impaired by drink, which is dispensed at numberless disreputable canteens among the mines. This vice is due in great measure to the indiscriminate granting of licenses, and to the callousness of those in whom the control is vested. The conditions under which licenses should be granted, and the whole question of natives and drink, are points to which the Chamber of Mines is now giving special attention.

The total number of natives employed in 1894 on the Witwatersrand goldfields was approximately 40,000, and the average of wages paid was 61s. per month of 28 days.[1] In addition to this, the cost of feeding natives on mealie meal and a small allowance of meat is about 8s. to 10s. per month. The returns for the first quarter of the present year show that the number of native employees has increased to 42,608.

Quarters, termed Compounds, are provided on every mine for the natives employed, and the mine staff generally includes a Compound manager, who is responsible for the maintenance of discipline, the sanitary condition of the Compound, and the welfare of the natives generally.

Table 5026, taken from the *Chamber of Mines Annual Report for* 1894, shows the distribution of whites and natives in the various departments of mining work.

[1] *Chamber of Mines Report*, 1894.

TABLE 5026.—DISTRIBUTION OF WHITE EMPLOYEES AT 31ST DECEMBER 1894.

Clerical.	Miners.	Mill Hands.	Cyanide Works.	Assayers, etc.	Contractors.	Fitters and Blacksmiths.	Engine Drivers.	Carpenters.	Various.	Total.	Total Amount Paid in Wages.	Remarks.
181	1968	344	213	62	395	399	476	418	907	5363	£1,396,959	From returns made by 67 companies.

DISTRIBUTION OF NATIVES.

Mine.	Fire-men.	Mill.	Cyanide.	Various Surface etc.	Total.	Average rate of Wages per month of four weeks, exclusive of food.	Remarks.
30,059	878	1436	1665	6850	40,888	£ 3 s. 1 d. 1	Compiled from returns made by 67 companies and estimate of number employed by other employers of labour in the mining industry. The average wages for unskilled labour are 58s. 9d., but this is increased by the higher wages paid to skilled labour.

WORKING COSTS.

Mining.—The item "mining," which appears in cost statements, generally comprises stoping and the conveyance of the ore to the shaft. In cases where hoisting and pumping are not given as a separate item, these are included in mining. Development costs are also sometimes included under the same head. The cost of mining is made up of wages paid to white miners, timbermen, and natives drilling and shovelling; a proportion of the pay of shift bosses and mine foremen, cost of explosives, detonators, fuses, candles, shovels, drill steel and hammers consumed, the wages of trammers and white overseers, and the maintenance of tram-lines and trucks. When machine drills are employed in stoping a proportion of compressor cost is added.

Native wages generally amount to from 40 to 45 per cent of the total cost, white labour to from 20 to 25 per cent, and explosives to from 12 to 20 per cent. The total cost varies very greatly according to the size of the ore-body mined, the condition of the walls, the hardness of the rock, to some extent on the scale of operations, and to a great extent on the management.

Mining and hauling costs are sometimes worked out to so much per ton on the basis of the tonnage hauled in the shafts. All other costs are reckoned on the basis of ore milled.

Hauling and Pumping.—Although sometimes included in statements of mining costs, these items are usually dealt with separately. The items of expenditure are wages of men in charge of loading stations, of banksmen, engine-drivers and stokers, and a proportion of the cost of supervision; the cost of material such as oil, waste, etc., and of fuel.

The total cost under this head is usually between 1s. 6d. and 3s. per ton.

Development.—The cost of the actual sinking of vertical shafts in "blue" ground, including raising and pumping, in a producing mine varies from £6 to £10 per foot. Timbering with sets and lagging in a large shaft may cost as much as £7 or £8 per foot, but usually it is somewhat less, and in shafts of smaller size, say 15 feet by 5 feet, it should not cost more than £3 or £4. In the preliminary opening up of deep level mines, where operations are confined to sinking, driving, etc., the cost of sinking, including timbering and underground equipment, sometimes amounts to £25 and even £30 per foot for depths of from 600 to 1000 feet. In such cases hauling and pumping, management, supervision, and general expenses are not distributed over a large tonnage, and consequently each foot of development has to bear a large proportion of the cost of these items. Inclined shafts can be sunk with greater speed and at less cost than vertical, owing to the face being more favourably conditioned for attack, and the greater facility for freeing the face of debris after blasting.

The cost of driving, raising, and sinking winzes in producing mines in blue ground is from £2 : 5s. to £3 : 5s. per foot, including track-laying. The cost of driving by hand labour is probably 15 to 25 per cent less than by machine drills. A proportion of the hauling and pumping costs is borne by development according to the time occupied in raising rock coming from the working ends.

The consumption of explosives in developing in blue ground amounts to from 4s. 6d. to 6s. 6d. per ton broken.

Sorting and Rock-breaking, where carried on together, cost from 8d. to 1s. per ton of sorted ore. Sorting averages about 7d. per ton of sorted ore.

Transport of Ore.—The cost of this varies with the distance the ore has to be carried and with the means employed. At the Langlaagte Estate it is about 3d. per ton by mechanical haulage. The mill is on the mynpacht and close to the mine, but the property is a large one, and there are three working shafts. At the Crown Reef the ore is transported about 1 mile by means of electric locomotives at a cost of about 3d. per ton. At the Van Ryan, where the distance is a mile or more and mules are employed, the cost is about 6d. per ton.

Transport seldom, under any existing conditions, amounts to more than 8d. per ton.

Milling.—The principal items of cost under this head are wages of millmen, amalgamators, and engine-drivers, fuel and maintenance. The last when it includes wear of shoes and dies, rock-breaker jaws, replacing of stems, cam-belts, etc., generally amounts in a fairly well-conducted mill to about 9d. per ton. The cost of milling depends more than anything else on the scale of operations and the efficiency of the plant. The cost of quicksilver not recovered is from ½d. to 1d. per ton crushed. The cost of battery water-supply is generally from 3d. to 6d. per ton crushed.

The total cost under this head varies from 2s. 6d. to as much as 7s. or 8s., but with the best mills it is seldom over 4s. per ton crushed.

Concentrating by Frue vanners varies from 4¾d. at the Langlaagte Estate, where the mill is large, to 7d. or 8d. per ton crushed with smaller plants.

Cyaniding. — The cost of cyaniding tailings depends chiefly on the size and nature of the plant, the latter being the more important factor. In well-designed plants, such as those at the Crown Reef, City and Suburban, Simmer and Jack, Robinson, etc., the cost of handling tailings in filling and discharging vats is considerably reduced. At the last-mentioned mine, where intermediate filling is in vogue, the cost of handling has been reduced to 10d. per ton. The cost of cyanide is from 1s. to 1s. 9d. per ton treated.

The total cost of cyaniding varies from 3s. 6d. to 10s. per ton treated, including royalty to the African Gold Recovery Company, at rates varying from 6 to 8 per cent of the value of gold recovered.

Office and General Expenses.—Under this head are included the following items of expenditure :—

Salaries in town and mine offices.

Directors' and auditors' fees.

Licenses, rents, insurances.

Printing, stationery, postage, advertising, etc.

Legal, medical, and engineering charges, etc.

Subscription to Hospital, Chamber of Mines, etc.

Interest and commissions.

Maintenance.—Expenditure under this head in maintaining the efficiency of machinery and underground equipment is always included in working costs, and is either recorded as a separate account, or is charged to the different accounts of mining, milling, transport, cyaniding and general charges, the latter undoubtedly being the better method.

Comparison of Working Costs.—In table 5027 will be found analyses of working costs at fifteen of the more important mines. It is a most difficult matter to compare the working costs of the companies upon a fixed basis, owing to the various systems of charging working expenditure to the different heads of charge: for instance, some companies charge all expenditure connected with mining operations to mining account, inclusive of expenditure upon drifting, rising, and sinking winzes, and maintenance of mining machinery and plant, whilst others charge drifting, rising, etc., to capital account, and redeem monthly according to the tonnage milled, charging the redemption to mine development redemption account, while maintenance of mining machinery, etc., is frequently charged to a special maintenance account. In this way some of the companies' mining expenses appear to be very low when compared with others. In order to arrive at a comparison of the actual mining costs it is necessary to add the following accounts appearing in the table, viz. mining, mine development redemption, rising and sinking, and where maintenance account appears separately, a proportion of its cost.

The same remarks apply in regard to milling costs; some companies charge maintenance of the mill to milling costs, while others charge this item direct to a special maintenance account.

The differences in the cost per ton under the head of general charges fluctuate to a great extent according to the milling capacity of the companies and to the various systems in vogue of charging items of expenditure thereto. Some

COMPARATIVE

GIVING DETAILS OF WORKING

Name of Company.	Year.	Period.	Number of Stamps.	Tons Milled.	Mining.	Mine Development and Redemption.	Raising and Pumping.
					s. d.	*s. d.*	*s. d.*
Worcester . . .	1894	July to Dec.	20	11,914	6 11.3 per ton mined 17,657	7 6	Included in mining
New Primrose . .	1893-94	July to June	100	157,236	8 11¼	3 7	1 5
Glencairn . . .	1893-94	19 months	50	74,706	7 10½	7 5½	1 5
Jumpers . . .	1895	...	100	...	8 10.27	Included	2 10.87
City and Suburban .	1895	February	120	...	13 3.08	4 6	Included in mining
New Chimes . .	1893-94	Nov. to Oct.	40	47,104	11 11.6	4 5.0	Included in mining
Henry Nourse . .	1893-94	July to June	25	23,417	*24 3.69	8 0	Included in mining
Meyer and Charlton .	1894	Jan. to Dec.	50	44,961	11 2	2 11.7	Included in mining
May Consolidated .	1894	Jan. to Dec.	50	31,048	12 2	4 2	Included in mining
Geldenhuis Estate .	1895	...	80	...	9 5½	6 0½	Included in mining
Robinson . . .	1894	Jan. to Dec.	Av. 64.5	107,935	14 3.76	8 4	Included in mining
Ferreira . . .	1894-95	July to Dec.	40	25,100	*26 0	11 6	Included in mining
Crown Reef . .	1894-95	April to Mar.	120	200,785	16 4.4	†1 3.0	Included in mining
Simmer and Jack .	1895	March	100	10,614	12 1.78	3 4	Included in mining
United Main Reef .	1895	March	50	6,120	17 6.64	Included in mining	Included in mining

N.B.—All costs given are per ton

By "Raising and Pumping" is meant the proportion of these costs

* These mining costs are greatly affected by the amount of waste sorted out ton mined was about 14s.

† This amount was written off cost of main shafts and

TABLE 5027,

COSTS AT SOME OF THE MINES.

Ore Transport.	Milling.	General Charges.	Maintenance	Depreciation.	Cyaniding including Royalty.	Tons Cyanided.	Remarks.
s. d.	s. d.	s. d.	s. d.	s. d.	s. d.		
0 3.7	6 1.5	5 4.8	7 3.6	6 7.6	...	...	Ore sorting 5½d. per ton.
0 8¾	2 6¼	1 1¼	Distributed	...	5 2	149,586	Cyanide costs per ton cyanided. Ore sorting included in transport.
0 4½	3 7	1 3	Distributed	1 10¾	5 3¾	40,222	Cyanide costs per ton cyanided. Mill water 1s. 6d. extra.
0 7.59	5 8.7	0 9.83	Distributed	...	...	...	...
Included in milling	4 8.27	1 4.44	Distributed	...	2 11.05	...	Cyanide costs per ton milled. Milling includes ore-sorting.
Included in milling	4 10.35	1 3.7	3 1.6	...	...	...	Milling includes concentrating 8.15d. per ton.
...	5 5.86	4 5.03	...	Included	10 1.45	21,800	Cyanide costs per ton cyanided. Ore-sorting and filling 2s. 7.62d. per ton.
0 5.24	3 5.79	4 1.05	2 7.8	1 0.08	5 9.2	22,900	Cyanide costs per ton cyanided.
0 9½	5 0	...	...	...	3 9	...	Cyanide costs per ton milled.
0 8¾	4 1¾	0 10	...	...	...	...	...
Included in milling	3 10.88	3 5.72	Distributed	2 0	3 10.6	70,526	Cyanide costs per ton cyanided. General charges include 7.17d. for maintenance of roads, etc.
0 4.7	5 9.0	Distributed	Distributed	...	...	...	...
0 2.72	3 0.3	2 7.3	Distributed	1 1.4	4 3.3	138,800	Cyanide costs per ton cyanided. Mining includes ordinary development.
0 7.28	5 2.12	3 9.0	...	...	...	...	...
Included in milling	3 7.36	0 8.35	Distributed	...	4 9.6	4,713	Cyanide costs per ton cyanided.

milled except where otherwise stated.
charged to "Mining" of ore. The rest is charged to "Development."
—the Ferreira probably mined upwards of 45,000 tons, so that the cost per
Similarly with "Mine Development."
cross-cuts. Drives and winzes are charged to "Mining."

companies charge the cost of the general management and the secretarial staff to this account, and others distribute this cost to driving, transport, milling, cyaniding, etc.

Mine Accounts.

The staff of a mine office includes a secretary or accountant at the head of affairs, general clerks, storekeepers, and timekeepers. Since the expenditure on material and in wages constitutes nine-tenths, at least, of the whole cost of working a mine, the remainder representing salaries paid for supervision, direction, office-work, etc., licenses and other general expenses, it naturally follows that *storekeeping and timekeeping* are of the first importance in mining accountancy.

The duties of the storekeeper and his assistants are:—(1) to take stock in detail of all goods received and stored, and of all articles issued, so that at all times a balance can be struck and a correct inventory returned of all material in stock; (2) to keep a record in detail of the distribution of all goods issued under heads of the various accounts, such as mining, development, transport, milling, etc. These are the main heads, but it is usual to go into greater detail, and to keep separate accounts for each level in the mine, for each shaft, and, in some cases, for individual stocks and so on, all of which is of the greatest use and importance in the intelligent management of a mine. It is usual to keep at least three books dealing with stores, viz. a stock-ledger into which is posted the quantity and value of all stores received and issued, a day-book or journal showing the quantity of stores issued, and a monthly summary of the rough day-book or journal, showing quantity and value of stores issued under the various headings of expenditure. The latter book, together with the invoices of stores received, is posted into the stock-ledger.

The timekeepers are responsible, in the first instance, for the record of time put in by white employees, and through them for natives' time. White men are paid at the end of each calendar month. Natives are paid on a weekly pay-day, but only on completion of twenty-four shifts, which, unless they have worked overtime, represents four weeks. In this way only a portion of the boys receive their wages on each pay-day. A white miner or ganger in charge of a number of boys is generally made responsible for the record of time put in by members of his gang, the timekeeper, while acting as a check, being responsible for an account of the distribution of labour employed in the various departments of work. As a check on the gangers and timekeepers it is customary to issue to each boy a cardboard or buckram ticket, on which are ruled-off spaces representing say twenty-four shifts. The native on completion of a shift has a space on his ticket marked off by his ganger, so that at the end of four weeks, or whenever all the spaces are marked, he can on presentation of his ticket on pay-days receive his wages. To simplify matters, it is usual to give each boy a number, which is written on his ticket, together with the name or number of his ganger. There is some variation in the details of systems of Kaffir timekeeping in vogue, but in a general way the above lines are followed. At the Simmer and Jack a native, besides carrying a monthly ticket on which his shifts are marked off, and on which a note is made of his ganger's number, the date of issue, and his own name and number, is given a daily ticket on which is written the ganger's number, his own number, and the number of the shift next succeeding on the monthly ticket. These daily tickets are supplied by the timekeeper to the gangers, who at the end of their shift return all surplus tickets to the timekeeper. The daily tickets are handed by the Kaffirs to the Compound manager, who keeps them as a check, and marks the shift on the monthly ticket. On the monthly ticket is stamped the rate of wages paid to the holder as fixed by his ganger in the presence of the timekeeper. This is perhaps the most elaborate system of Kaffir timekeeping practised on the Rand.

Since native pay-days never coincide with the end of a calendar month, it will be readily understood that in drawing up monthly statements of cost an allowance has to be made for native wages earned but not paid during the month. For the purpose of securing accuracy in the monthly statements, an account of such wages due is kept under the head of "Native Wages Suspense Account," the shifts not paid for being recorded at an average rate obtained from the rates of wages actually paid during the month.

On all the larger mines close attention is now paid to the accurate and systematic record of all the items which make up the cost of the various operations, so that if desired an analysis can be made down to the minutest detail. Very complete information of this kind is contained in the published reports of the Robinson, Crown Reef, Simmer and Jack, Geldenhuis Estate, and many other companies. Monthly statements of working expenditure and revenue are published by nearly all producing companies.

Profit and loss accounts generally show the expenditure and sums written off under some or all of the following heads: Mining, hauling and pumping, transport, breaking and sorting of ore, milling and concentrating, cyaniding, maintenance, mine development, depreciation of plant, buildings, etc., general charges and dividends. As an illustration of the methods of rendering such statements, we will instance the last annual report of the Crown Reef, where for a number of years past very systematic accounts have been kept under the direction of Mr. F. Raleigh, to whom we are indebted for much useful information on this head.

The greatest variations in accounts occur in connection with maintenance, mine development, and depreciation.

Maintenance is, as a rule, fully charged as working

expenditure, but in some cases it is distributed over the various departments as mine maintenance, mill maintenance, general maintenance, etc., while in others all costs of maintaining machinery, permanent plant, shafts, and so on, are stated summarily in one account.

Mine development is dealt with in various ways. It is usual to keep separate accounts of moneys expended in permanent mine works, such as shafts and cross-cuts, and in development works as generally understood, *i.e.* drives and winzes. The amounts periodically written off from the former account and debited to working expenditure are always more or less arbitrary. At the Crown Reef the whole cost of sinking and equipping main shafts from one level to another is charged to the ore between these levels, and redeemed by an estimated amount added to the cost of mining each ton extracted. At the Robinson the whole amount expended in shafts in any month is written off at the end of that month. Many companies, however, are not so liberal, especially if the balance to credit is small. It is an extremely difficult thing to fix on an equitable course in the matter of writing off main shafts, for although the shaft generally progresses in accordance with the requirements of the mine, the part first sunk is as useful and valuable in working the lowest as in working the upper levels, and the life of a mine is always an element of great uncertainty owing to a variety of circumstances, among which is the continual decrease in working costs, and consequent increase of the amount of workable ore reserves, owing to reefs becoming payable that were not so at the outset.

In the matter of drives and winzes, although the same difficulties obtain to a certain extent, more definite lines are followed. On most mines the same method of writing off such work is adopted as is in use at the Crown Reef in the matter of main shafts, *i.e.* the amount expended in opening each level is distributed over the estimated tonnage of the then pay-ore developed by them, a proportionate amount being thus arrived at which is chargeable to each ton of ore as it is extracted. The amount appears in monthly statements of working costs under the head of Redemption of Mine Development.[1] On some mines the total amount expended on development works, *i.e.* on drives and winzes, during each month is charged to working costs for that month, irrespective of the relation of the quantity developed to the quantity actually mined. This is done at the Crown Reef, Robinson, Jubilee, and two or three other mines.

The amounts written off for *depreciation* of plant, buildings, etc., are variable and somewhat arbitrary. Generally speaking, the more prosperous a company is the more regular, systematic, and liberal it is in respect of sums written off. When old inefficient plant is about to be discarded it is in most cases written down to actual sale value.

The Robinson Company for the year 1894 wrote off 5 per cent from all plant, buildings, etc., the total depreciation amounting to £10,573. In the latter half of the year 1894 the Ferreira Company wrote off from 20 to 25 per cent from machinery and plant, buildings and permanent works, a total amount of £30,000. The Worcester G. M. Company for the same period wrote off as follows:—

Machinery joint account (*i.e.* Ferreira Worcester Joint Mill, etc.)	5 per cent.
Machinery, plant, and tools	10 "
Permanent works	10 "
Buildings	10 "
Permanent waterworks	10 "
Live stock, carts, etc.	5 "
Furniture	5 "

The Crown Reef Company, for the year ending 31st March 1895, wrote off a total amount of £23,761.

[1] Redemption of mine development has thus usually a somewhat different signification from that obtaining in the Crown Reef accounts.

REDUCTIONS IN WORKING COSTS.

Future reductions in working costs may be looked for chiefly in the following directions :—

1. Reduced price of dynamite.
2. Reduced price of fuel.
3. Reduced wages of natives.
4. Daily increasing experience, and consequent improvement in the general economy of mining, and other operations.
5. Growth in the scale of operations carried on by individual companies, and consequent distribution over larger tonnages of all general costs.

Reduction in the Price of Dynamite.—The present tax on dynamite amounts to about 37s. 6d. per case, *i.e.* over 40 per cent of the present price. As with little doubt state-craft in the Transvaal will advance and improve in the course of time, a very considerable reduction of this tax may be anticipated; say it is reduced, as it would be in most civilised countries, to a maximum of 7s. 6d. per case. This would mean a reduction in the price of dynamite of 33 per cent. Now the cost of explosives constitutes 12-per cent of the total working costs in mining and all associated operations. Therefore we may anticipate in time, perhaps not a very great number of years, a reduction of 4 per cent in this direction.

Reduction in the Price of Fuel.—About one-half of the price of coal at the present time is due to transport charges. When the conditions of transport will be ameliorated is very doubtful, but that some future reduction of the present high rates may be anticipated is reasonable. It is also reasonable to suppose that some reduction in the cost of winning coal will result from a reduction in natives' wages and improved conditions of mining generally. Moreover, if the probabilities of the above reductions be granted, competition would further affect the market. Taking coal at 18s. per ton about the present average, we think it probable that before many years this price will be reduced to 13s. or 14s. per ton, a reduction of say 25 per cent.

Fuel constitutes about 9 per cent of total working costs. Therefore we may anticipate in this direction a 2 per cent reduction.

Reduction in Native Wages.—The wages paid to natives at the present time, at the rate of 60s. a head per month, constitute at least 32 per cent of total working costs. Prognostications as to this element in costs must be extremely uncertain both as to time and amount, but there can be no doubt that the fight, even if a prolonged one, will be continued with unremitted zeal on the part of those interested in the industry, and it is, in our opinion, equally certain that some measure of success will eventually be obtained. We have already dealt with the measures that are being taken, and need not refer further to the matter here. The rate ultimately hoped for by the Chamber of Mines is 40s. a month, but if half this reduction is obtained, or say 17 per cent, this would mean over 5 per cent less in total working costs.

Reductions due to Improvements in General Economy and Increased Scale of Operations.—These, although perhaps the most certain factors in future reduction of costs, are indefinite and difficult to assess. The latter is the more important of the two, and will probably in the course of a few years affect costs to the extent of at least 1s. per ton, or say 3 per cent.

To sum up, we have a total anticipated reduction of 14 per cent on present working costs. The average total costs at the mines now working is 29s. 6d., say 30s., per ton; 14 per cent of this is 5s., leaving 25s., which we think will within the next lustrum prove to be the average working cost of outcrop properties.

CHAPTER XI

GOLD LAW AND MINING REGULATIONS

Gold Law

It is hardly within the scope of the present volume to go into any great detail regarding the mining laws of the South African Republic, but the following *résumé* of the main points of "The Gold Law," as set forth in Act No. 14 of 1894, may be of general interest.

Article 1 of this law reads as follows:—

"The right of mining for and disposing of all precious metals and precious stones belongs to the State."

Article 5 reads thus:—

"His Honour, the State President, has the power, with advice and consent of the Executive Council, to proclaim and throw open Government grounds, and after consultation with the owner, if possible, also private grounds, as public fields."

Notice of such proclamation is published in the *Government Gazette* during six weeks, and further posted at the office either of the Mining Commissioner or the Landdrost of the district.

If, however, the owner of private grounds does not prospect, nor give permission to do so to others, the Government has not the right to throw open the ground as a public digging.

The discoverer of precious metals in payable quantities on private farms or on Government grounds, at least to 6 miles distant from an already worked locality, is entitled on proclamation of such grounds to select six claims,[1] called prospecting claims, of which he becomes the owner.

The owner of private ground can, prior to proclamation, allot to any person or persons a number of claims not exceeding sixty, called "vergunning" claims. The owner himself is entitled to "owners' claims" according to the following scale:—For a piece of ground 50 morgen[2] or less in extent one claim, for a piece of ground 200 morgen in extent two claims, and for every additional 250 morgen in extent one claim, with a maximum of ten claims for a farm. Prospectors' claims have prior rights to owners' claims.

The owner of private proclaimed ground receives half the licenses paid by claim-holders on his ground.

The owner has the further right to mark off for himself, for mining purposes, one-tenth of the ground to be proclaimed, this portion being held by him under mining lease, or "mynpacht-brief," for a term of not less than five nor more than twenty years, with right of renewal. The rental on mynpachts is 10s. per morgen per annum, but the Government has the right at all times to demand in lieu of this rent 2½ per cent of the value of the finds of the past year.

The proportion of the breadth of a mynpacht to its length along the strike of the reefs must not be more than one to two.

The owner, besides owners' claims and mynpacht, can retain a certain area for occupation, *i.e.* for residential and farming purposes, termed "werf," and he has the first right to sufficient water for both agricultural and mining purposes on the areas retained by him.

[1] A reef claim is 150 Cape feet wide, measured in the direction of the strike of the reef, by 400 Cape feet in the direction at right angles. In English measurement a claim is 155 feet by 413 feet, equal to about 1½ acres, or more exactly 63,984 square feet.

[2] A morgen is rather more than two English acres (1 morgen = 2.117 acres).

On the proclamation of any farm or piece of ground as a public digging, claims can be pegged out by any law-abiding white person, male or female, on their having obtained a license to peg, one license being granted to each person. Any person using another's license must, within one month of registration, file that person's power of attorney with the duplicate license in the office of the Mining Commissioner of the district.[1]

The amount due on a "prospecting" license on private grounds is 5s. per month per claim, on Government grounds 2s. 6d. Such a license is converted into a "digger's" license at discretion of the Mining Commissioner, when the amount due per claim is increased to 20s. per month. If, however, no quartz is being extracted, or there is no machinery on the property, the amount of a "digger's license" is limited to 15s. per month.

If claim-license monies are not paid on the date on which they are due, thirty days' grace is given, during which the holder can renew his license. If such renewal does not take place within the first fourteen of such thirty days of grace a fine is inflicted. If the licenses are not renewed the Government may sell the claims by public auction, or restore them to their owners on payment of arrears of licenses and fines.

The risk of losing claims by one month's neglect of licenses can be obviated by special registration, obtainable under specified conditions, which we need not detail. Claims

[1] The number of claims registered in the books of the Government Mining Commissioners on the 31st December 1894 was 55,302, being distributed as follows:—

Witwatersrand district	46,002
Heidelberg district	4,965
Klerksdorp district	3,997
Venterskroon district	338
Total	55,302

State Mining Engineer's Report of the year 1894.

"specially registered" do not finally revert to the Government until the licenses are nine months overdue. On specially registered claims mortgages may be given as on fixed property.

The proprietary right ("Bezitrecht") of holders of claims, water-rights, tramways, machine-stands, and tailing-sites, is decided by the Mining Commissioner, after calling for objections during three months in the *Government Gazette*, appeal to the High Court or Circuit Court being allowed during the three months following the decision of the Mining Commissioner.

"Bezitrecht" once granted is indisputable.

Water rights are vested in the State, and must be applied for to the Mining Commissioner, in whose discretion the granting of them lies.

The Bewaarplaats Question.[1]—Bewaarplaats, or depositing sites for tailings, etc., are granted by the Mining Commissioner on application. A bewaarplaats is 200 feet square in area, and is held under a monthly license of 2s. 6d. It is only a surface right, and is usually granted on ground believed to be non-metalliferous. The bewaarplaats license gives the holder of it no right to the minerals which may exist beneath the ground held by virtue of such license. In the early days of the Witwatersrand goldfields bewaarplaats were in many cases granted to companies on ground which was afterwards found to constitute the deep levels of the Main Reef, and consequently of great mineral wealth. On this discovery numerous applications were made to the Government for the underground mining rights of these properties. The Government was given the right by the Volksraad to accede to these applications on rules and regulations which were to be framed. These rules and regulations have, however, not yet been made, owing no

[1] With regard to the Bewaarplaatsen question we have consulted Mr. C. A. Wentzel, of Messrs. Dumat and Wentzel, Johannesburg, who has given us the above information.

doubt to the great dispute which immediately arose, and has since continued for a considerable time, as to whom these underground mining rights should be granted. The companies claim that they, as holders of the surface rights and of the adjoining mining rights, should have the preference. Their opponents, on the other hand, claim that the rights should be granted to any persons making application in the order of priority. At present the question stands over for settlement by resolution of the Volksraad.

Mining Regulations.

The mining regulations, of which the following is a condensed abstract, were passed in June 1893. Their administration belongs to the department of the State Mining Engineer, and the mines are under the direct supervision of the mining inspectors, of whom there are three for the Witwatersrand district. The jurisdiction lies in the first instance with the Courts of Landrosts (Magistrates) and Mining Commissioners, the right of appeal in the ordinary way being granted. Should an accident be attributable to intention or gross carelessness, the case is treated in terms of the ordinary criminal procedure.

Section 1.—Protection of the Surface.

All open works, cuttings, shafts, etc., out of use must be protected by fences and notice boards.

Safety pillars must be left under all public places and works.

Water, whether run to waste or reserved, that contains poisonous chemicals, must be closed in.

Section 2.—Safety of Mining Works.

Boundaries of mining properties underground must be protected by a 10-foot pillar, to prevent the flooding of one mine by the water from another.

Shafts where unsafe must be timbered.

The timbering of pack-walling of working stopes where unsafe is enjoined.

Section 3.—Safety of Miners.

Shafts, winzes, ore-shoots, etc., must be railed in or protected by a movable cover.

Efficient and readily-handled brakes on engines are compulsory.

All hauling-shafts must be supplied with proper signalling appliances, for signalling between the surface and all levels. And in underground haulages every point must be adapted for signalling the engine-driver. Refuge cuttings must be made at intervals of not less than 50 yards.

A code of signals is specified for use in all mines.

In ladderways, in steeply inclined or vertical shafts, resting-places must be introduced every 30 feet. Ladders must not incline backwards, and can only be fixed vertically with the consent of the mining inspector. Ladders must project 3 feet above every resting-place or landing, and hand-rails must be provided. Ladderways must be securely partitioned off from hoisting or other compartments in a shaft.

Every mine must have at least two exits approachable from all parts of the mine.

Hauling and hoisting ropes must have a tensile strength equal to six times the load, and a reserve rope must be kept on the mine.

The safety of all hauling and hoisting appliances must be tested weekly.

Depth indicators in the engine-room at main shafts are compulsory.

The use of safety clutches is enjoined.

Here follow certain regulations of no great importance to gold mines regarding ventilation and lighting.

Magazines for explosives must be erected 200 yards from dwellings and public places. If more than 5000 lbs. of explosives are stored the magazine must be enclosed by an earth or stone bank as high as the roof. There must be a lightning conductor. Nitro-glycerine preparations must not be stored in the same compartment as gunpowder, and detonators must be kept separate.

Underground magazines must be 200 feet from working drives and shafts, and must not contain more than 500 lbs. of any kind of explosive. Not more than one day's supply may be served out for any working-place, and the distribution is to be in the hands of duly appointed persons.

Coloured persons are not allowed to prepare cartridges, load holes, or blast.

Then follow blasting regulations similar to those of the British code.

Boilers and driving appliances must be under the supervision of a qualified person.

All dangerous parts of machinery must be railed in.

Section 4.—*Workmen.*

Every mine must keep a register of all persons working above or underground.

Changing rooms must be provided.

Section 5.—*Measures to be adopted in case of accident.*

The regulations under this head are made chiefly with a view to the facilitation of due inquiry into the causes of accidents, and to the provision of immediate relief to those who may have suffered injury.

Section 6.—*Mining Plans.*

The drawing and maintaining of accurate detailed plans of the surface and of underground works is enjoined on all mining companies. The plans must be drawn according to the true meridian. Mine plans must be posted every six months. All bench marks and fixed survey points must be shown, also strike and dip of reefs, etc., and faults and dykes. All general plans must be drawn to a scale of 1 to 5000; mine plans may be on a scale of 1 to 500, or of 1 to 1000. All underground workings about to be abandoned must be surveyed before they become inaccessible. Misrepresentation in plans is subject to a penalty of £500 or one year's imprisonment.

Section 7.—Registration of Returns from Mines and Metallurgical Works.

It is compulsory to send in to the Government representatives monthly reports of the gold won, the progress of mine works, and the number of employees on surface and underground.

Section 8.—Mine Work on Sundays and Holidays is prohibited except work which, if stopped, might cause complete cessation.

Section 9.—Special Regulations may be formed by individual mining companies or managers provided they do not clash with the Government regulations.

Section 11.—Sundry Regulations.

No one intoxicated or seriously ill may enter a mine.

Government officials have free access to mines at all times, and are allowed to take samples of the ore or minerals if they deem it advisable in the execution of their duty.

The mining regulations must be posted in conspicuous places at the mines.

Section 12.—The Supervision of Mines.

One qualified manager must be the responsible representative of a mine.

A manager must be certified by a Board of Examiners

who use their discretion in granting certificates. Two years' practical experience, or the approval of the State Mining Engineer, is necessary.

Penalties, varying from £5 to £500, with the alternative of imprisonment, may be incurred by the contravention of mining regulations.

ACCIDENTS.

The following extract from the *Report of the State Mining Engineer for the year* 1894 shows the nature and number of accidents occurring in Witwatersrand gold mines. It is to be hoped that the strict enforcement of the mining regulations by the mining inspectors will tend to further reduce the number of accidents due to negligence or carelessness on the part of those in authority:—

STATEMENT OF ACCIDENTS

WHICH OCCURRED IN THE MINES OF THE SOUTH AFRICAN REPUBLIC DURING THE YEAR 1894, AS REPORTED TO THE INSPECTORS OF MINES.

Cause of Accidents.	Witwatersrand Gold Mines.						Heidelberg Gold Mines.						Klerksdorp Gold Mines.					
	White.			Coloured.			White.			Coloured.			White.			Coloured.		
		Inj.			Inj.			Inj.			Inj.			Inj.			Inj.	
	Killed.	Serious.	Slight.	Killed.	Serious.	Slight.	Killed.	Serious.	Slight.	Killed.	Serious.	Slight.	Killed.	Serious.	Slight.	Killed.	Serious.	Slight.
Drilling into misfired holes	...	2	...	17	5	21	...	...	...	...	...	...	...	...	...	2	6	5
Other dynamite explosions	5	6	1	34	17	37	...	...	...	1	...	2	1	1	...	3	...	1
Machinery	1	2	5	12	3	5	...	...	...	...	...	...	...	...	...	2	...	...
Falls of rock (in slopes, etc.)	...	3	1	39	16	17	...	...	...	...	...	...	...	...	...	1	1	1
Falling materials, etc. (in shafts, etc.)	...	2	1	6	4	6	...	...	...	...	...	...	...	...	...	...	...	...
Trucks and tramming	...	1	...	3	10	10	...	...	...	1	...	...	...	...	...	...	2	...
Travelling by cage or skip	9	5	5	35	16	11	...	...	...	...	...	...	...	...	...	1	...	...
,, ,, ladders	...	...	2	4	1	2	...	...	...	...	...	...	...	...	...	...	...	...
Falling in shafts	3	2	4	14	4	3	...	...	...	...	1	...	...	...	...	...	...	1
Miscellaneous	3	2	1	6	3	5	...	...	...	...	...	...	...	...	...	...	...	...
Totals	21	25	20	170	79	117	...	...	...	2	1	2	1	1	...	9	9	8
Number of cases of accidents	321						4						16					

CHAPTER XII

GOLD PRODUCTION AND DIVIDENDS

IN this chapter we propose to give some statistics bearing on the gold production and dividends paid by the mining companies. In compiling the tables, which form an important adjunct to this chapter, we have had recourse to the excellent monthly returns published by the Witwatersrand Chamber of Mines. It is necessary to point out that the returns made by the Chamber are of bullion, and not of fine gold, but as the cash value of the bullion is also given, they can, if required, be converted to fine or to standard gold. In most cases the value stated represents the actual price paid by the banks for the purchase of the bullion; but where this is not known average values are taken. The average value of mill gold is about 72s. 6d. per oz., and of cyanide gold about 60s. per ounce.

Annual Outputs of the Mines.—Table I. gives the annual output of each mine in the Witwatersrand fields since the start of the mills in 1887 up to the present date.

The yield from each source, viz., mill, concentrate, and tailings, is returned separately, so that by means of this table the performance of any mine can be exactly traced.

The mines are arranged in the order of their performance. At the head of the list comes the Robinson, with a total of 749,636 oz.; followed by the Langlaagte Estate, with 581,189 oz.; the Crown Reef, 433,089 oz.; the Primrose, 326,868 oz.; the Simmer and Jack, 252,703 oz.; the Ferreira, 250,947 oz.;

TABLE I—ANNUAL OUTPUT OF THE MINES IN THE WITWATERSRAND GOLDFIELDS FROM 1887 TO JUNE 1895.

Name of Company.	1887.	1888.	1889.	1890.	1891.		1892.			1893.			1894.			1895. 6 months.			Total Mill.	Total Concentrates.	Total Tailings.	Total Ounces.	Approximate Total Value.[1]
					Mill.	Tailings.	Mill.	Concentrates.	Tailings.	Mill.	Concentrates.	Tailings.	Mill.	Concentrates.	Tailings.	Mill.	Concentrates.	Tailings.					
Robinson	...	26,287	75,038	71,823	74,759	19,196	98,799	8,397	37,594	104,223	10,660	17,921	110,962	14,217	20,213	52,716	7,741	11,672	614,607	41,015	106,596	762,218	£2,697,285
Langlaagte Estate	...	9,998	60,855	56,750	51,462	...	69,980	2,318	16,684	65,813	9,047	30,050	86,661	12,388	47,320	48,457	9,579	16,835	449,976	33,332	110,889	594,197	2,082,984
Crown Reef	2,788	11,019	18,364	37,500	39,841	...	50,726	...	19,487	51,688	723	29,679	73,339	4,458	42,749	40,703	4,599	16,687	325,967	9,780	108,602	444,349	1,541,962
Primrose	...	2,028	3,626	12,800	23,078	...	45,391	...	11,084	57,574	...	26,204	67,447	397	21,177	48,462	152	19,464	260,406	549	77,929	338,884	1,180,615
Simmer and Jack	...	2,471	12,550	21,262	38,409	...	38,140	...	508	38,905	...	767	45,475	1,297	14,953	29,628	1,800	15,154	226,840	3,097	31,382	261,319	927,762
Ferreira	414	137	220	...	46,340	2,978	35,342	...	12,890	43,978	...	11,697	52,547	...	13,499	27,486	3,765	7,484	206,464	3,765	48,548	258,777	907,136
City and Suburban	1,106	5,194	18,814	13,072	12,410	...	29,125	...	...	37,778	...	9,044	49,823	...	25,442	38,819	199	19,188	206,141	199	53,674	260,014	910,083
Durban Roodepoort	...	3,839	14,539	16,018	23,156	...	30,931	...	3,324	37,883	...	22,751	38,448	...	22,562	22,331	...	11,144	187,145	...	59,781	246,926	857,643
Jumpers	977	10,455	21,778	30,231	16,918	...	19,305	...	...	38,635	294	...	47,663	...	14,819	24,823	3,106	7,125	210,785	3,400	21,944	236,129	842,419
Geldenhuis Estate	507	4,599	1,239	...	31,123	...	47,727	...	...	47,434	...	...	43,636	535	14,742	22,428	495	14,360	198,693	1,028	29,102	228,823	810,629
Nigel	...	1,506	3,685	13,601	10,191	3,565	26,571	3,031	20,685	25,455	3,516	17,036	28,089	...	23,460	12,786	...	10,646	121,884	6,547	75,392	203,823	691,907
May Consolidated	...	2,048	9,949	13,088	27,491	...	27,690	696	...	24,957	...	2,875	25,356	...	13,750	16,618	...	11,317	147,197	696	27,942	175,835	621,615
Wemmer	6,104	6,266	8,384	7,017	13,959	...	16,522	...	...	22,707	...	3,063	34,024	...	9,151	24,149	3,051	7,848	139,132	3,051	20,062	162,245	576,104
Meyer and Charlton	1,200	6,571	4,971	15,557	20,617	...	19,137	...	3,195	27,329	...	6,854	25,583	...	8,276	13,309	...	3,892	134,274	...	22,217	156,491	553,513
Jubilee	1,608	4,937	6,406	12,333	8,786	501	11,577	208	5,534	24,774	36	5,255	21,564	...	5,511	11,860	...	2,631	103,845	244	19,432	123,521	435,343
Salisbury	1,092	11,527	9,304	8,508	11,336	...	15,335	...	5,490	19,269	...	5,587	14,615	...	6,108	8,769	...	2,718	99,755	...	19,903	119,658	420,898
Langlaagte Block B	...	...	...	...	14,689	...	24,368	...	...	19,621	100	6,870	21,846	2,557	11,238	13,224	1,734	3,986	93,748	4,391	22,094	120,233	421,748
Glencairn	...	...	...	867	5,690	...	19,305	...	...	15,914	...	...	26,091	859	20,925	18,972	340	9,178	86,839	1,199	30,103	118,141	409,487
Chimes	...	...	5,238	1,527	7,783	...	28,383	...	...	14,510	...	7,297	23,800	...	4,884	12,585	351	2,544	93,826	351	14,725	108,902	385,524
Worcester	...	5,555	8,041	11,472	11,735	...	16,696	...	...	17,122	...	...	27,067	...	...	10,835	1,412	...	108,523	1,412	...	109,935	398,295
Heriot	...	4,922	7,816	5,796	247	...	...	...	...	...	...	8,690	35,259	...	14,236	22,846	...	11,656	76,886	...	34,582	111,468	382,644
United Main Reef	...	2,092	1,076	11,052	7,598	...	14,242	...	...	9,929	...	...	21,194	...	13,895	19,411	...	7,230	86,595	...	21,125	107,720	377,189
Randfontein	...	...	...	...	85	...	6,586	...	...	23,311	...	6,624	34,806	411	6,915	19,964	1,105	3,257	84,752	1,516	16,796	103,064	363,383
Champ d'Or	...	...	...	...	7,298	...	16,990	...	...	5,722	...	1,778	30,341	...	14,367	15,795	...	7,686	76,146	...	23,831	99,977	347,516
Langlaagte Royal	...	...	...	1,375	8,776	...	18,147	...	...	16,477	...	...	21,098	...	19,014	8,447	...	6,059	74,320	...	25,073	99,393	344,810
Henry Nourse	...	4,046	4,327	4,882	5,810	...	8,879	...	...	15,330	...	6,893	20,898	...	8,237	11,290	...	3,941	75,462	...	19,071	94,533	330,762
Wolhuter	307	2,717	1,548	684	...	...	11,093	...	...	25,049	...	...	21,039	...	9,558	10,085	...	4,478	72,522	...	14,036	86,558	305,106
Stanhope	1,144	4,158	6,886	6,385	10,909	...	14,125	...	3,874	10,790	...	6,042	10,215	...	...	4,353	...	2,182	68,963	...	12,098	81,061	286,476
George Goch	855	379	1,103	5,150	10,701	...	10,523	...	...	7,181	...	...	18,285	...	9,355	11,474	150	8,242	65,651	150	17,597	83,398	291,277
Johannesburg Pioneer	...	4,970	5,765	9,542	9,261	...	11,672	...	191	9,073	...	1,120	10,952	...	...	6,616	...	...	67,851	...	1,311	69,162	249,730
New Rietfontein	...	...	...	...	...	...	11,597	...	...	28,169	80	6,958	15,397	...	6,275	9,922	104	3,687	65,085	184	16,920	82,189	287,491
Orion	...	...	1,319	...	3,248	...	7,108	...	3,896	8,318	...	9,678	14,748	...	19,461	6,385	...	11,948	41,126	...	44,983	86,109	283,996
Van Ryn	...	1,220	750	2,199	...	...	8,907	...	...	13,440	...	...	17,303	...	3,846	12,135	491	5,650	55,954	491	9,496	65,941	233,140
Paarl Central	...	...	509	8,731	2,283	...	6,681	...	...	...	...	958	17,716	...	7,005	11,284	...	4,209	47,204	...	12,172	59,376	207,740
Village	...	...	...	...	...	...	6,974	...	1,159	6,144	...	1,996	26,077	...	4,813	743	...	605	39,938	...	8,573	48,511	170,494
Geldenhuis Main Reef	...	...	739	7,291	3,606	...	5,195	...	...	3,417	...	373	9,917	...	6,076	8,118	...	3,804	38,283	...	10,250	48,533	169,271
Princess	...	...	...	...	8,765	...	8,375	...	...	756	...	...	14,116	...	3,900	8,475	...	2,919	40,487	...	6,819	47,306	167,354
Kleinfontein	...	...	1,982	260	4,471	...	...	...	...	4,779	...	...	15,018	...	4,569	11,509	...	4,428	38,019	...	8,997	47,016	164,809
Aurora	...	1,742	2,319	3,589	6,894	...	9,588	...	...	9,109	...	...	3,425	...	...	938	...	...	37,604	...	...	37,604	136,315
Treasury	578	4,234	1,415	...	3,768	...	6,992	...	...	7,587	...	4,285	5,177	...	2,835	...	...	...	29,751	...	7,120	36,871	129,207
Metropolitan	110	1,091	850	1,389	600	...	7,652	...	...	402	23	...	12,066	...	1,476	8,554	...	4,361	32,714	23	5,837	38,574	136,345
Langlaagte United	...	...	...	5,666	4,968	...	3,863	...	...	10,621	...	...	4,819	...	...	2,058	...	916	31,995	...	916	32,911	118,724
New Unified	...	576	...	1,507	825	...	4,645	...	...	16,484	...	...	4,589	...	...	...	...	...	28,626	...	...	28,626	103,815
Main Reef	...	2,717	2,209	...	...	...	12,764	...	...	2,589	...	...	6,189	...	2,453	...	...	...	26,468	...	2,453	28,921	103,305
New Aurora West	...	...	...	...	...	...	12,614	...	...	10,351	...	...	4,089	...	1,197	...	...	...	27,054	...	1,197	28,251	101,662
Roodepoort (Kimberley)	...	1,583	1,668	7,909	5,737	...	6,042	...	...	981	...	...	1,066	...	...	...	...	...	24,986	...	...	24,986	90,574
Witwatersrand	...	...	4,450	2,009	4,276	...	1,820	...	...	10,315	...	...	...	...	...	...	...	...	22,870	...	...	22,870	82,904
May Deep[2]	...	...	...	591	5,751	...	7,471	...	...	5,392	...	...	2,876	...	284	...	...	...	22,081	...	284	22,365	80,896
Meyer and Leeb	...	...	1,477	2,329	1,928	...	4,400	...	...	3,971	...	...	5,599	20	541	2,105	...	...	21,809	20	541	22,370	80,749
Star	...	...	...	...	921	...	8,221	...	...	4,324	1,040	3,516	...	...	4,559	...	...	...	13,466	1,040	8,075	22,581	76,809
Moss Rose[3]	...	4,839	2,180	3,124	9,709	...	831	...	...	220	...	...	...	...	...	...	...	...	20,903	...	...	20,903	75,773
George and May (Botha's Reef)	...	...	1,011	2,389	3,335	...	684	...	...	2,042	...	...	5,812	...	3,137	...	...	3,689	15,273	...	6,826	22,099	75,842
Spes Bona	...	162	...	1,861	4,743	...	...	...	...	...	...	...	6,544	...	...	...	...	...	13,310	...	...	13,310	48,252
Crœsus	...	214	3,798	4,557	4,047	...	...	...	...	...	...	...	...	...	...	2,060	...	...	14,676	...	...	14,676	53,149
Langlaagte Western[4]	...	...	...	707	5,038	...	4,508	...	...	2,061	...	...	...	...	...	...	...	...	12,314	...	...	12,314	44,638
Banket	...	...	1,051	456	1,778	...	2,609	...	...	1,348	...	...	1,214	...	1,682	1,588	...	1,590	10,044	...	3,272	13,316	46,327
Evelyn	...	720	1,392	2,045	4,958	...	...	...	...	...	...	2,546	...	...	...	...	...	...	9,115	...	2,546	11,661	40,680
Gipsy	...	...	...	1,399	1,068	...	3,604	...	...	1,563	...	713	786	...	2,177	...	...	...	8,420	...	2,890	11,310	39,192
Knight's Tribute	...	...	...	...	...	...	...	...	...	2,127	...	2,356	4,583	...	...	...	...	...	6,710	...	2,356	9,066	31,395
Nabob	...	540	...	2,270	3,661	...	1,446	...	...	...	...	...	...	...	...	...	...	...	7,917	...	...	7,917	28,699
Modderfontein	...	...	...	884	...	...	532	...	...	2,997	...	...	3,404	...	...	...	...	...	7,817	...	...	7,817	28,335
Tharsis	...	...	1,512	4,414	217	...	...	...	...	...	...	...	...	...	...	...	...	...	6,143	...	...	6,143	22,269
Mint	...	3,978	1,700	...	...	...	...	...	...	...	...	...	...	...	...	...	...	...	5,678	...	...	5,678	20,583
Percy[5]	...	3,740	798	906	...	...	...	...	...	...	...	...	...	...	...	...	...	...	5,444	...	...	5,444	19,735
Angle Tharsis	...	...	...	1,677	...	...	659	...	...	2,636	...	...	...	...	...	...	...	...	4,972	...	...	4,972	18,023
Moss Rose Extension	...	...	3,703	1,154	...	...	...	...	...	...	...	...	...	...	...	...	...	...	4,857	...	...	4,857	17,607
Grahamstown[6]	...	...	3,043	1,177	...	...	...	...	...	...	...	...	...	...	...	...	...	...	4,220	...	...	4,220	15,298
Steyn Estate	...	...	3,266	298	583	...	...	...	...	...	...	...	...	...	...	...	...	...	4,147	...	...	4,147	15,036
Gardner Main Reef	...	...	...	...	2,357	...	783	...	...	...	...	...	...	...	...	...	...	...	3,140	...	...	3,140	11,386
Vogelstruis	...	...	...	...	1,287	...	1,820	...	...	...	...	...	...	...	...	96	...	...	3,203	...	...	3,203	11,607
Wolsingham	...	...	2,987	...	...	...	...	...	...	...	...	...	...	...	...	...	...	...	2,987	...	...	2,987	10,828
Emma	...	...	...	...	2,522	...	...	...	...	...	...	...	...	...	...	...	...	...	2,522	...	...	2,522	9,142
Blue Sky	...	...	...	...	525	...	834	...	...	...	...	...	1,055	...	...	...	...	...	2,414	...	...	2,414	8,751
Bantjes	...	1,347	818	...	...	...	...	...	...	...	...	...	...	...	...	...	...	...	2,165	...	...	2,165	7,848
Ethel	...	...	...	1,511	374	...	...	...	...	...	...	...	...	...	...	...	...	...	1,885	...	...	1,885	6,833
Stanhope Geldenhuis	...	...	24	...	925	...	...	...	...	...	...	...	...	...	...	...	...	...	949	...	...	949	3,440
Edinburgh	...	...	...	...	863	...	...	...	...	...	...	...	...	...	...	...	...	...	863	...	...	863	3,129
Dupreez	...	...	743	...	...	...	...	...	...	...	...	...	...	...	...	...	...	...	743	...	...	743	2,693
Doornfontein Mint	...	...	...	128	...	...	...	...	...	...	...	...	...	...	...	...	...	...	128	...	...	128	464
	18,790	166,424	359,235	462,717	656,489	26,240	942,526	14,650	145,595	1,020,574	25,519	267,476	1,291,708	37,139	512,642	715,211	40,172	296,407	5,633,674	117,480	1,248,360	6,999,514	£24,594,621

[1] Mill and concentrate gold at 72s 6d per ounce, tailings gold at 60s per ounce
[2] Amalgamated with New Primrose
[3] Do do
[4] Amalgamated with Crœsus
[5] Do Treasury
[6] Do Jubilee

[*Face p.* 280

The material originally positioned here is too large for reproduction in this reissue. A PDF can be downloaded from the web address given on page iv of this book, by clicking on 'Resources Available'.

the City and Suburban, 250,551 oz.; the Durban Roodepoort, 241,072 oz.; the Jumpers, 229,108 oz.; and the Geldenhuis Estate, 221,210 oz.

The annual totals of this table can be summarised as follows :—

	Mill.	Concentrates.	Tailings.	Total Oz.
	Oz.	Oz.	Oz.	
1887	18,790	...	...	18,790
1888	166,424	...	...	166,424
1889	359,235	...	...	359,235
1890	462,717	...	...	462,717
1891	656,489	...	26,240	682,729
1892	942,526	14,650	145,595	1,102,771
1893	1,020,574	25,519	267,476	1,313,569
1894	1,291,708	37,139	512,642	1,841,489
1895 6 months	715,211	40,172	296,407	1,051,790
	5,633,674	117,484	1,248,360	6,999,514

The grand total is 6,999,514 oz. of a value of £24,594,621.

Monthly Outputs of the Mines.—Table II. gives the total monthly outputs of the Witwatersrand goldfields. Under this head the Heidelberg district is included, but not the Klerksdorp or Potschefstroom district, this being in accordance with the practice of the Johannesburg Chamber of Mines.

The totals for each year are slightly in excess of those given in Table I. There are two reasons for the difference.

1. The Chamber of Mines' returns, from which this table has been compiled, include the alluvial gold found in these districts.

2. The totals include an item of "gold from undeclared sources."

Gold Production.—On the Production Chart will be found a graphic representation of the production from the start of the

II.—TABLE OF MONTHLY OUTPUTS (IN

	1887.	1888.	1889.	1890.	1891.	
					From Mills, including Concentrates.	From Tailings.
	Oz.	Oz.	Oz.	Oz.	Oz.	Oz.
January	...	7,030.5	25,505.12	35,006.15	52,595.8	610.0
February	...	12,189.14	22,456.18	36,887.5	48,532.2	1,547.0
March	...	11,975.15	27,919.0	37,780.2	51,349.1	1,600.0
April	...	14,726.8	27,028.16	38,696.19	54,726.16	1,645.0
May	887.3	13,446.13	35,028.7	38,836.5	53,612.1	1,061.0
June	734.0	12,770.5	30,877.13	37,419.10	54,263.15	1,600.0
July	240.0	15,686.17	31,091.2	39,456.14	52,750.7	2,174.3
August	1,408.15	18,815.19	30,519.14	42,863.11	56,051.4	3,019.0
September . . .	1,935.19	20,242.4	34,143.10	45,485.19	62,412.15	3,189.0
October	4,028.10	27,165.6	32,214.6	45,248.17	69,202.17	3,590.11
November . . .	5,457.3	26,826.17	33,721.16	46,782.18	68,289.15	5,104.0
December . . .	8,457.8	26,784.6	39,050.11	50,352.5	71,980.4	8,332.7
Totals . . .	23,148.18	207,660.9	369,557.5	494,817.0	695,766.5	33,472.1
Value . . .	£81,022	£726,821	£1,300,509	£1,735,491	£2,445,161	£111,167

SUMMARY.

Year.	Yield in Ounces.			Value.		Total.
	As per Table.	Undeclared Sources.	Total.	As per Table.	Undeclared Sources.	
	Oz.	Oz.	Oz.	£	£	£
1887	23,149	...	23,149	81,022	...	81,022
1888	207,660	...	207,660	726,821	...	726,821
1889	369,557	...	369,557	1,300,509	...	1,300,509
1890	494,817	...	494,817	1,735,491	...	1,735,491
1891	729,238	...	729,238	2,556,328	...	2,556,328
1892	1,173,817	35,111	1,208,928	4,166,851	123,882	4,290,733

OUNCES) OF THE WITWATERSRAND MINES.

1892.		1893.		1894.		1895.	
From Mills, including Concentrates.	From Tailings.	From Mills, including Concentrates.	From Tailings.	From Mills, including Concentrates.	From Tailings.	From Mills, including Concentrates.	From Tailings.
Oz.	Oz.	Oz.	Oz.	Oz.	Oz.	Oz.	Oz.
72,567.9	9,692.2	89,886.18	15,160.13	105,054.4	39,888.14	125,187.11	50,119.17
74,424.14	10,366.2	74,597.5	15,539.0	106,563.7	41,269.1	117,915.4	48,770.11
80,223.1	11,473.4	89,696.3	18,877.18	116,035.16	44,664.6	128,886.12	53,436.1
80,581.0	11,260.15	89,722.10	18,707.10	117,333.16	47,147.0	130,089.5	52,099.7
84,589.5	10,933.1	91,287.10	21,278.9	115,466.4	49,143.15	135,893.16	55,750.11
85,749.9	15,722.11	94,993.9	23,373.4	115,040.2	48,800.1	143,229.14	54,446.8
82,367.13	15,042.8	95,850.5	25,860.6	115,917.18	48,241.5	...	...
85,252.8	14,208.12	99,966.18	29,913.9	116,997.19	53,926.13	...	...
88,717.7	15,263.18	94,167.14	30,544.4	119,645.14	52,527.6	...	...
93,562.15	16,013.3	97,501.14	32,574.12	116,915.14	53,630.12	...	...
87,448.19	15,346.2	97,817.11	35,217.8	118,357.10	54,898.0	...	...
98,140.1	14,871.13	103,631.14	37,451.16	126,660.5	53,252.1	...	...
1,013,624.1	160,193.11	1,119,119.11	304,498.9	1,389,988.9	587,388.14	781,802.2	314,622.15
£3,664,443	£502,408	£4,061,185	£933,870	£5,026,839	£1,772.472	£2,825,147	£954,799

SUMMARY—*continued.*

Year.	Yield in Ounces.			Value.		Total.
	As per Table.	Undeclared Sources.	Total.	As per Table.	Undeclared Sources.	
	Oz.	Oz.	Oz.	£	£	£
1893	1,423,618	52,884	1,476,502	4,995,055	185,035	5.180,090
1894	1,977,377	45,821	2,023,198	6,799,311	160,311	6,959,622
1895 (6 mos.)	1,096,425	25,336	1,121,761	3,779,946	59,977	3,839.923
	7.495,658	159,152	7,654,810	2,6141,334	529.205	26,670,539

mills in 1887 up to the end of 1894. It shows the monthly returns of—

(1) The total gold produced (in bullion ounces).

(2) The tonnage of ore crushed.

(3) The tonnage of tailings treated.

(4) The average mill-grade (in dwts).

It will be seen that, apart from small temporary fluctuations, the lines indicate steady and uniform increases in returns both of ore milled and gold produced.

For the first four years (May 1887 to June 1891), a period embracing the infancy of the industry, and naturally one of trial and uncertainty, the increase in production was slow and subject to fluctuations. During that period the output rose to 55,000 ounces, representing on the average an increase of about 13,000 ounces per annum. In July 1891 a rapid expansion in gold production began and continued to June 1892, the monthly output rising during that period from 55,000 ounces to 101,000 ounces, being an increment of nearly 100 per cent during the twelve months. After ten months of irregular movements, which advanced the monthly output by only a few thousand ounces, another period of rapid expansion set in. The upward movement began in May 1893, and continued uninterruptedly, with the exception of a small set-back in September 1893, to the end of September 1894—a period of 17 months, during which the monthly output rose from 108,000 to 172,000 ounces, being an increment of over 60 per cent. The rise has continued at a slower rate with intermittent fluctuations, and in June of the present year the record figure of 200,941 ounces was attained.

Summary of Dividends paid by Witwatersrand Gold Mining Companies.—Table III. gives the amount of each dividend paid by every Witwatersrand Gold Mining Company since 1887.[1]

[1] We have to express our acknowledgments to Messrs H. Eckstein and

TABLE III.—SUMMARY OF DIVIDENDS PAID BY WITWATERSRAND GOLD MINING COMPANIES.

Name of Company.	1887.			1888.			1889.		
	No. of Dividends.	Per Cent per Year.	Cash Distribution.	No. of Dividends.	Per Cent per Year.	Cash Distribution.	No. of Dividends.	Per Cent per Year.	Cash Distribution.
			£			£			£
Jumpers	1	5	2,100	...	...	...	2-3	20	9,400
Jubilee	1-5	25	3,375	6-12	45	7,950	13-17	25	6,250
Wemmer	1-4	57	15,960	5-7	15	4,700	8-9	15	4,950
Moss Rose	...	...	...	1-4	20	7,400	...	...	...
Royal (Treasury)	...	...	...	1-2	17½	3,500	...	...	...
Worcester	...	...	...	1-2	30	4,500	3-5	45	9,080
Crown Reef	...	...	...	1-2	12	12,000	3-4	12	12,000
Salisbury	...	...	...	1-9	117½	23,900	10-14	82½	19,275
Geldenhuis Estate	...	...	...	1	5	4,000	...	...	...
Stanhope	...	...	...	1-4	22½	6,750	5-7	20	6,800
City and Suburban	...	...	...	1	5	2,960	2-7	50	31,951
Meyer and Charlton	...	...	...	1-2	20	8,300	...	...	...
Moss Rose Extension	...	...	...	1	4	4,000	...	...	...
Aurora	...	...	...	1	5	2,500	...	...	...
Grahamstown	...	...	...	1	10	1,650	...	...	...
Golden Kopje	...	...	...	1	5	3,750	2	2½	1,875
Percy	...	...	...	1	10	1,000	2	10	1,000
Evelyn	...	...	...	1	10	2,600	...	...	...
Paarl Pretoria	...	...	...	...	...	...	1	25	18,750
Simmer and Jack	...	...	...	...	...	...	1	10	7,500
New Heriot	...	...	...	...	...	...	1-3	15	8,995
Langlaagte Estate	...	...	...	...	...	...	1-4	21	129,690
Durban Roodepoort	...	...	...	...	...	...	1	20	18,000
Chimes (New)	...	...	...	...	...	...	1-2	10	4,000
Robinson	...	...	...	...	...	...	1	5	135,937
Johannesburg Pioneer	...	...	...	...	...	...	...	...	...
New Primrose	...	...	...	...	...	...	...	...	...
Ferreira	...	...	...	...	...	...	...	...	...
Champ d'Or	...	...	...	...	...	...	...	...	...
Langlaagte Royal	...	...	...	...	...	...	...	...	...
New Crœsus	...	...	...	...	...	...	...	...	...
Orion	...	...	...	...	...	...	...	...	...
Nigel	...	...	...	...	...	...	...	...	...
New Aurora West	...	...	...	...	...	...	...	...	...
Meyer and Leeb	...	...	...	...	...	...	...	...	...
Langlaagte Block B	...	...	...	...	...	...	...	...	...
Rietfontein	...	...	...	...	...	...	...	...	...
Kleinfontein	...	...	...	...	...	...	...	...	...
Roodepoort United Main	...	...	...	...	...	...	...	...	...
Glencairn	...	...	...	...	...	...	...	...	...
May Consolidated	...	...	...	...	...	...	...	...	...
Geldenhuis Main	...	...	...	...	...	...	...	...	...
	...	...	21,435	...	...	101,460	...	...	425,453

Name of Company.	1890.			1891.			1892.		
	No. of Dividends.	Per Cent per Year.	Cash Distribution.	No. of Dividends.	Per Cent per Year.	Cash Distribution.	No. of Dividends.	Per Cent per Year.	Cash Distribution.
			£			£			£
Jumpers	...	...	...	...	...	...	...	...	...
Jubilee	18-28	55	14,300	29-33	35	9,100	34-41	60	17,502
Wemmer	...	...	...	10	10	4,630	...	...	...
Moss Rose	...	...	...	...	...	...	...	...	...
Royal (Treasury)	...	...	...	...	...	...	...	...	...
Worcester	6-8	15	13,108	9-10	10	9,071	11-13	17½	15,155
Crown Reef	5	15	15,900	6-7	50	53,000	8-9	55	63,000
Salisbury	...	...	...	...	...	...	15	30	27,900
Geldenhuis Estate	...	...	...	2	15	18,000	3	10	17,500
Stanhope	...	...	...	8-12	70	23,800	...	...	...
City and Suburban	8-10	75	58,750	11-12	50	42,500	...	...	...
Meyer and Charlton	3-5	20	12,648	6-9	50	31,620	10-11	45	29,320
Moss Rose Extension	...	...	...	...	...	...	...	...	...
Aurora	...	...	...	...	...	...	2	5	2,600
Grahamstown	...	...	...	...	...	...	...	...	...
Golden Kopje	...	...	...	...	...	...	...	...	...
Percy	...	...	...	...	...	...	...	...	...
Evelyn	...	...	...	...	...	...	...	...	...
Paarl Pretoria	...	...	...	...	...	...	...	...	...
Simmer and Jack	...	...	...	2-4	30	24,900	5-8	40	33,400
New Heriot	...	...	...	...	...	...	...	...	...
Langlaagte Estate	5-6	15	67,050	7-8	10	44,700	9-10	20	89,400
Durban Roodepoort	2-4	30	30,000	5-6	20	20,000	7-9	30	35,833
Chimes (New)	...	...	...	...	...	...	1-7	40	20,049
Robinson	...	...	...	2	4	108,750	3-5	12	326,250
Johannesburg Pioneer	1	10	2,100	2-4	35	7,350	5-8	50	10,500
New Primrose	...	...	...	1-3	15	26,044	4-6	20	34,726
Ferreira	...	...	...	1	50	22,000	2-3	125	56,250
Champ d'Or	...	...	...	1	10	8,010	...	...	...
Langlaagte Royal	...	...	...	1	5	2,500	2-7	30	17,000
New Crœsus	...	...	...	...	...	...	1-2	10	7,500
Orion	...	...	...	...	...	...	1-2	7½	1,725
Nigel	...	...	...	...	...	...	1-3	47½	75,986
New Aurora West	...	...	...	...	...	...	1-2	10	6,000
Meyer and Leeb	...	...	...	...	...	...	1-3	17½	2,575
Langlaagte Block B	...	...	...	...	...	...	...	...	...
Rietfontein	...	...	...	...	...	...	...	...	...
Kleinfontein	...	...	...	...	...	...	...	...	...
Roodepoort United Main	...	...	...	...	...	...	...	...	...
Glencairn	...	...	...	...	...	...	...	...	...
May Consolidated	...	...	...	...	...	...	...	...	...
Geldenhuis Main	...	...	...	...	...	...	...	...	...
	...	...	213,856	...	...	455,975	...	...	890,171

Name of Company.	1893.			1894.			1895 (to June).			Total amount Distributed.
	No. of Dividends.	Per Cent per Year.	Cash Distribution.	No. of Dividends.	Per Cent per Year.	Cash Distribution.	No. of Dividends.	Per Cent per Half Year.	Cash Distribution.	
			£			£			£ s.	£ s.
Jumpers	4-5	25	25,000	6-7	55	55,000	...	...	...	91,500 0
Jubilee	42-44	90	27,000	45-49	150	45,000	50-51	60	30,000 0	160,477 0
Wemmer	...	...	...	...	...	...	11	100	55,000 0	85,240 0
Moss Rose	...	...	...	...	...	...	...	...	...	7,400 0
Royal (Treasury)	...	...	...	...	...	...	...	...	...	3,500 0
Worcester	14-16	15	13,605	17-20	50	45,359	21-22	35	31,751 0	141,629 0
Crown Reef	10-11	50	60,000	12-13	50	60,000	14	25	30,000 0	305,900 0
Salisbury	16	10	9,300	...	...	...	...	...	...	80,375 0
Geldenhuis Estate	4-5	25	43,750	6-7	30	56,250	8	30	60,000 0	199,500 0
Stanhope	13	50	17,000	14-15	100	34,000	16-17	30	10,200 0	98,550 0
City and Suburban	...	...	...	...	...	...	13-14	75	63,750 0	199,911 0
Meyer and Charlton	12-13	60	43,012	14-15	55	42,869	1	25	21,250 0	189,019 0
Moss Rose Extension	...	...	...	...	...	...	...	...	...	4,000 0
Aurora	...	...	...	...	...	...	...	...	...	5,100 0
Grahamstown	...	...	...	...	...	...	...	...	...	1,650 0
Golden Kopje	...	...	...	...	...	...	...	...	...	5,625 0
Percy	...	...	...	...	...	...	...	...	...	2,000 0
Evelyn	...	...	...	...	...	...	...	...	...	2,600 0
Paarl Pretoria	...	...	...	...	...	...	...	...	...	18,750 0
Simmer and Jack	9-12	40	34,000	13-14	20	17,000	15-16	20	50,000 0	166,800 0
New Heriot	4	10	8,500	5-6	40	34,750	7	75	63,750 0	115,995 0
Langlaagte Estate	11-12	30	140,100	13-16	45	200,150	17	25	117,500 0	788,590 0
Durban Roodepoort	10-13	55	68,750	14-17	60	75,000	18-19	30	37,500 0	285,083 0
Chimes (New)	...	...	...	8-9	15	15,000	10-11	40	40,000 0	79,049 0
Robinson	6	4	108,750	7-9	14	380,650	10	6	165,000 0	1,225,337 0
Johannesburg Pioneer	9-11	37½	7,875	...	...	...	...	...	...	27,825 0
New Primrose	7-9	47½	91,591	10-11	40	141,830	12	25	70,000 0	364,191 0
Ferreira	4-6	135	60,750	7-8	150	89,500	9	65	58,500 0	287,000 0
Champ d'Or	...	...	...	2	20	16,020	3	20	25,698 0	49,728 0
Langlaagte Royal	8	5	7,000	...	...	...	...	...	...	26,500 0
New Crœsus	...	...	...	...	...	...	...	...	...	7,500 0
Orion	...	...	...	3-12	105	31,500	13	10	12,880 0	46,105 0
Nigel	4-7	50	79,985	8-11	50	79,985	12-13	20	31,994 0	267,950 0
New Aurora West	3	5	4,000	...	...	...	...	...	...	10,000 0
Meyer and Leeb	4	10	1,500	...	...	...	...	...	...	4,075 0
Langlaagte Block B	1	8	5,694	2	8	5,694	3	8	5,694 0	17,082 0
Rietfontein	1-2	25	40,000	...	...	...	...	...	...	40,000 0
Kleinfontein	...	...	...	1	12½	13,750	2	12½	13,750 0	27,500 0
Roodepoort United Main	...	...	...	1-2	20	25,000	3	25	32,500 0	57,500 0
Glencairn	...	...	...	1-2	12½	25,000	3	15	33,750 0	58,750 0
May Consolidated	...	...	...	...	...	...	1	10	26,500 0	26,500 0
Geldenhuis Main	...	...	...	...	...	...	1	5	7,500 0	7,500 0
	...	...	897,162	...	...	1,489,307	...	...	1,094,467 0	5,589,286 0

[Face p. 286.

The material originally positioned here is too large for reproduction in this reissue. A PDF can be downloaded from the web address given on page iv of this book, by clicking on 'Resources Available'.

MONTHLY PRODUCTION-CHART.

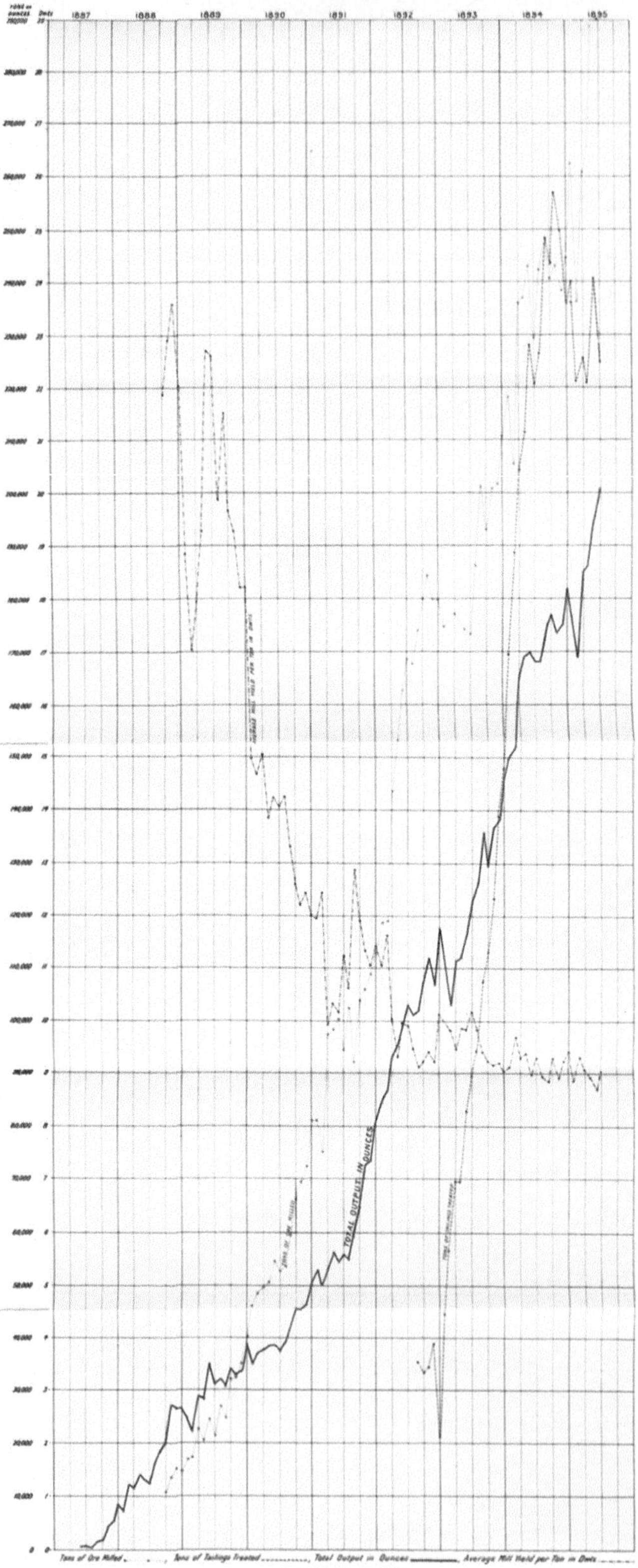

COMPILED BY F H HATCH

Face p 286

To accompany F H HATCH and J A CHALMERS' "The Gold Mines of the Rand"

The material originally positioned here is too large for reproduction in this reissue. A PDF can be downloaded from the web address given on page iv of this book, by clicking on 'Resources Available'.

The amounts paid each year are as follows:—

1887	£21,435	0	0
1888	101,460	0	0
1889	425,453	0	0
1890	213,856	0	0
1891	455,975	0	0
1892	890,171	0	0
1893	897,162	0	0
1894	1,489,307	0	0
1895 (six months)	1,094,467	0	0
Total to date	£5,589,286	0	0

The most regular dividend-payer has been the Jubilee, which, beginning in 1887, has made an uninterrupted series of distributions down to the present year, the total number of payments having been 51, equal to 455 per cent of the capital, regard being had to changes in capitalisation. Next in order come the Worcester and the Crown, both of which have paid regularly since 1888, the former having made 22 payments, amounting to 217½ per cent of the capital, and the latter 14, amounting to 269 per cent. The largest total distribution is that of the Robinson Company, amounting to £1,225,337 in 15 payments, equal to 45 per cent of the capital. Next comes the Langlaagte Estate, which has distributed a total amount of £788,590 in 17 payments, equal to 166 per cent of the capital. The Durban Roodepoort has paid a total of £285,083 in 30 payments, equal to 251 per cent of the capital. The City and Suburban has paid a total amount of £199,911 in 14 payments, equal to 255 per cent of the capital. The Simmer and Jack, £166,800 in 16 payments, equal to 160 per cent. The Meyer and Charlton, £167,769 in 15 payments, equal to 250 per cent. The Geldenhuis Estate, £199,500 in 8 payments, equal to 115 per cent. The Nigel, £267,950 in 13 payments, equal to

Company for their courtesy in allowing us access to their books, by means of which we have been enabled to compile this table.

167½ per cent; and the Ferreira £287,000 in 9 payments, equal to 525 per cent of the capital.

Table IV. (p. 292) is a schedule of mining companies quoted on the Johannesburg Stock Exchange.[1]

It shows the nominal capital, issued capital, and the market valuation as at June 19 of this year, of—

(*a*) Main Reef outcrop properties.
(*b*) Deep Level properties.
(*c*) Black Reef properties, and those in outlying districts.

A number of the Deep Level companies, notably the offshoots of the Rand Mines, are not quoted on the Johannesburg Stock Exchange, and, consequently, do not appear in this list. Financial corporations, such as the Consolidated Goldfields and East Rand Proprietary, are of course excluded.

According to this table the total nominal capital of the Witwatersrand Mines listed therein amounts to £20,140,100, of which £18,984,377 have been issued. The value placed by the market on these stocks on June 19, 1895, amounted to £81,970,107. As the Deep Level companies not included in the list have a market value amounting to at least £20,000,000, it appears that the total valuation of Rand mining stocks amounts to over £100,000,000 sterling.

Future Production.—In order to make a rough estimate of the probable future production of the fields, it is necessary to consider what additions are contemplated to the stamping power. On Table, p. 295, we have tabulated the number of stamps now running (June), and those that we anticipate will be dropping at the beginning of the next century.

The number of stamps at work at the end of 1894 was 2169. By June the number has increased to 2642. A very large increase is contemplated within the next few years, and it will be seen by our table that our estimate of the number of

[1] For this list we are indebted to Messrs. Foxwell and Roberts of Johannesburg.

stamps that will be running 5 years hence amounts in round numbers to 8000. The present average crushing capacity is rather over 4 tons per stamp per day; but this figure is increasing daily, and it is probable that in 5 years' time the average will be 5 tons, owing principally to the greatly increased weight of new stamps,[1] and partly, no doubt, to the use of screening of coarser mesh, which is daily gaining favour. Taking an average of 335 days' milling in the year, we get 13,400,000 tons milled annually, say 13,000,000. With a total extraction of 10 dwts. fine gold per ton (which is a little under the present average), this would yield 6,500,000 ounces, of a total value £26,000,000. The total yield for 1894 (Chamber of Mines), was 2,024,000 ounces, valued at £6,963,000.

From reports made in the early part of this year by companies[2] working between the Princess (Roodeport) and the Glencairn (Driefontein) inclusive we have deduced the following averages:—

From 207,000 tons milled.

Average yield per ton 45s. 6d., equal to about 12 dwts. fine gold.

Average working cost, 29s. per ton, including depreciation.

Average profit per ton, 16s. 6d.

From our knowledge of the mines we estimate that an average thickness of reef of 6 feet is being worked.

[1] See page 195.

[2] The companies actually figured on are the following:—Princess, Roodepoort United Main Reef, Durban Roodepoort, Langlaagte Block B, Paarl Central, Langlaagte Estate, Crown Reef, Robinson, Worcester, Ferreira, Wemmer, Salisbury, Jubilee, City and Suburban, Meyer and Charlton, Wolhuter, George Goch, Metropolitan, Henry Nourse, Heriot, Jumpers, Geldenhuis Estate, Simmer and Jack, New Primrose, May Consolidated, Glencairn. The companies within the same limits, from which actual figures are not at present obtainable, *i.e.* from Roodepoort to Driefontein inclusive, are the Banket, Evelyn, Kimberley Roodepoort, Vogelstruis, Bantjes, Odessa, Aurora West, Aurora, Unified, Main Reef, Nabob, Tharsis, Star, Crœsus, Langlaagte, Agnes Munro, Cinderella, Blue Sky, St. Angelo, Balmoral, Driefontein, Ginsberg.

Taking a mean dip below present workings of 30°, and making a liberal deduction for faulted ground and other short-falls, the average profit to be derived from each claim of working outcrop properties will be on the above data £22,000.

The producing properties within the limits mentioned cover a length of reef of 13 miles, the companies at present (June 1, 1895) non-producing cover 14 miles.

The average obtained for the central portion of the Rand from the Langlaagte Estate to the Glencairn, a distance of 11 miles, in which practically every mine is at the present time earning profits, are:—

From 188,000 tons milled,—

Average yield per ton 46s. 4d., equal about 12 dwts. fine gold.

Average working costs, 28s., including depreciation.

Average profit per ton, 17s. 4d.

Average thickness of reef worked, about 6 feet.

Average profit per claim (less allowance for short-falls) £23,000.

In order to arrive at some estimate of the total production that may be expected from the main section of the Rand as above limited, from Roodepoort to Driefontein, 27 miles, it is necessary to assume an average yield from the mines that are not at present working. Nearly all of these mines have been worked to a greater or less extent, and without doubt they will be worked again, in many cases with a good result, in some few with perhaps only a small margin of profit, or possibly a loss. For our estimates we will assume an average total extraction of 9 dwts. fine gold, equal to 38s. per ton, and let us also assume that a thickness of only 3 feet[1] of reef will be worked. The section in question mined to 8000 feet from the outcrop on the slope

[1] In many of these mines only one reef and portions of another are payable.

of the reef (a vertical depth of say 3500 feet) would yield with say 25 per cent deduction for faults, etc., including 5 per cent for portions already worked out, 290 million tons, the gold from which would equal, at an average of, say, 10¼ dwts. per ton, 148 million ounces of a value of 592 millions sterling.

Taking the central portion of the Rand from Langlaagte Estate to the Glencairn, a length of 11 miles, we have for 8000 feet of backs, with a 20 per cent reduction for short falls of reef, including 5 per cent for the portion already worked out, and with 6 feet thickness of reef, 170 million tons at, say, 45s. per ton, equal to £382,000,000, or about 95 million ounces.

With the reductions in working costs anticipated (see Economics, page 271) it seems to us unlikely that the average cost of winning and treating this ore will exceed the cost at the present day, or say, 30s. per ton. If we assume this figure, then, the cost of obtaining 95,000,000 ounces of gold would be £255,000,000, and the profit derivable from the reefs to the depth assumed would be £127,000,000.

Without going into any close estimates or predictions with regard to the remaining portions of the Main Reef Series and the outlying portions of the district, we think we may safely forecast a production from the Witwatersrand generally, within the next half century, of upwards of £700,000,000 in value, of which in all probability £200,000,000 will be clear profit.

TABLE IV.

Schedule of Mining Companies, showing Capitalisation and Market Valuation at 19th June 1895. Johannesburg Quotations.[1]

(*a*) *Main Reef Outcrop Properties.*

Name of Company.	Nominal Capital.	Issued Capital.	Prices. 19th June.	Market Valuation. 19th June.
Agnes Munro . . .	£100,000	£93,000	16/3	£75,562
Aurora	65,000	65,000	26/-	84,500
Aurora West United . .	140,000	100,000	*35/-	175,000
Balmoral	100,000	100,000	57/6	287,500
Banket	300,000	280,000	5/-	45,000
Bantjes Reef . . .	95,000	83,000	67/-	278,050
Bohemian . . .	131,000	131,000	4/3	27,837
Champ d'Or . . .	135,000	120,000	*75/-	450,000
Cinderella . . .	100,000	100,000	26/6	132,500
City and Suburban . .	85,000	85,000	*500/-	2,125,000
Consolidated Angle-Tharsis	80,000	65,740	42/-	138,054
Crown Reef . . .	120,000	120,000	*200/-	1,200,000
Driefontein . . .	130,000	120,000	27/-	162,000
East Roodepoort . .	30,000	29,000	*15/-	21,750
Evelyn	80,000	66,000	*20/-	66,000
Ferreira	90,000	89,000	*350/-	1,557,500
Geldenhuis Estate . .	200,000	200,000	136/-	1,360,000
Geldenhuis Main Reef .	150,000	150,000	20/-	150,000
George and May . .	150,000	140,000	*25/-	175,000
George Goch Amalgamated	100,000	100,000	63/6	317,500
Ginsberg	160,000	152,000	*35/-	266,000
Glencairn . . .	225,000	225,000	81/-	911,250

* Approximate Price.

1 The following off-shoots of the Rand Mines are not quoted in Johannesburg :—

Company.	Nominal Capital.	Issued Capital.	Company.	Nominal Capital.	Issued Capital.
Rose Deep . .	£400,000	£300,000	Nourse Deep . .	£450,000	£375,000
Geldenhuis Deep .	350,000	280,000	Crown Deep . .	300,000	250,000
Jumpers Deep . .	400,000	300,000			

Also the following recent flotations :—

Company.	Nominal Capital.	Company.	Nominal Capital.
Simmer West	£400,000	West Kleinfontein . . .	£160,000
Simmer East	700,000	Kleinfontein Deep . . .	100,000
Glen Deep	600,000	Modderfontein B. Extension	325,000
Knight's Deep	550,000	Glenluce (Knight's Tribute)	225,000
South Salmon	525,000	Violet Consolidated . .	500,000
Nigel Deep	500,000	Horsham Monitor . .	190,000
Central Nigel Deep . . .	200,000	Rip (Fern)	65,000
Western Nigel . . .	300,000	Lancaster	300,000
Transvaal Nigel . . .	250,000	East Orion	250,000
Rand Nigel	120,000	Pleiades	175,000
West Nigel Deep . . .	60,000	Black Reef Proprietary .	150,000
Ryan Nigel	105,000	Buffelsdoorn A . . .	250,000
Van Ryn West . . .	170,000	United Buffelsdoorn Mines .	250,000
North Van Ryn . . .	170,000	Southleigh	225,000
Chimes West . . .	200,000	Eastleigh Deep . . .	300,000

TABLE IV.—*Continued.*

(*a*) *Main Reef Outcrop Properties*—Continued.

Name of Company.	Nominal Capital.	Issued Capital.	Prices. 19th June.	Market Valuation. 19th June.
Henry Nourse	£125,000	£125,000	*120/-	£750,000
Johannesburg Pioneer	21,000	21,000	110/-	115,500
Jubilee	50,000	50,000	*200/-	500,000
Jumpers	100,000	100,000	117/6	587,500
Langlaagte Block B.	632,000	606,174	*22/6	681,946
Langlaagte Estate	470,000	467,000	105/-	2,451,750
Langlaagte Royal	150,000	150,000	*70/-	525,000
Langlaagte Star	200,000	170,000	*30/-	255,000
Langlaagte Unified	150,000	150,000	55/-	412,500
Luipaard's Vlei Estate	350,000	248,000	*17/6	217,000
Main Reef	112,000	112,000	27/-	151,200
May Consolidated	275,000	265,500	65/-	862,875
Metropolitan	100,000	100,000	51/6	257,500
Meyer and Charlton	85,000	83,200	150/-	624,000
Modderfontein	200,000	200,000	*320/-	3,200,000
Nabob	75,000	71,500	*8/-	28,600
New Blue Sky	150,000	150,000	11/	82,500
New Chimes	100,000	100,000	62/6	312,500
New Comet	225,000	175,000	62/6	546,875
New Crœsus	500,000	500,000	54/-	1,350,000
New Gipsy	45,000	45,000	5/9	12,937
New Grahamstown	150,000	150,000	21/-	157,500
New Heriot	85,000	85,000	225/-	956,250
New Kleinfontein	100,000	100,000	100/-	500,000
New Primrose	280,000	280,000	134/-	1,876,000
New Spes Bona	160,000	160,000	53/-	424,000
New Unified	160,000	145,000	13/-	94,250
Paarl Central	200,000	200,000	24/6	245,000
Princess	160,000	142,000	57/6	408,250
Randfontein Estate	2,000,000	2,000,000	43/-	4,300,000
Robinson	2,750,000	2,750,000	*180/-	24,750,000
Roodepoort Durban	135,000	125,000	*140/-	875,000
,, Kimberley	125,000	125,000	*75/-	468,750
,, United Main Reef	130,000	130,000	123/-	799,500
Ruby	30,000	23,114	*220/-	254,254
Salisbury	100,000	93,000	97/6	453,375
Simmer and Jack	250,000	250,000	330/-	4,125,000
Stanhope	35,000	34,000	*40/-	68,000
Teutonia	100,000	90,000	*20/-	90,000
Treasury	60,000	60,000	65/-	195,000
Van Ryn Estate	160,000	105,000	185/-	971,250
Vogelstruisfontein	45,000	41,800	5/6	11,495
Wemmer	55,000	55,000	232/6	639,375
Witwatersrand	250,000	250,000	172/6	2,156,250
Wolhuter	130,000	130,000	140/-	910,000
Worcester	100,000	90,727	*85/-	385,589
Westleigh	250,000	225,000	14/6	163,125
York	120,000	90,000	20/-	90,000
	£15,001,000	£14,332,755	...	£70,298,399

* Approximate Price.

TABLE IV.—*Continued.*

(*b*) *Deep Level Properties.*

Name of Company.	Nominal Capital.	Issued Capital.	Prices. 19th June.	Market Valuation. 19th June.
Bonanza	£200,000	£200,000	30/-	£300,000
Champ d'Or Deep	275,000	166,000	42/6	352,750
Langlaagte Block B Deep Level	75,000	75,000	16/6	61,875
Paarl Ophir	12,000	12,000	*120/-	72,000
Roodepoort Deep Level	180,000	170,000	*80/-	680,000
,, Durban Deep	350,000	290,000	82/6	1,196,250
Rietfontein Deep	275,000	250,000	16/-	200,000
Village Main Reef	180,000	180,000	130/-	1,170,000
	£1,547,000	£1,343,000	...	4,032,875

* Approximate Price.

(*c*) *Properties on Black Reef, Kimberley Series, etc., and those in Outlying Districts.*

Name of Company.	Nominal Capital.	Issued Capital.	Prices. 19th June.	Market Valuation. 19th June.
Amazon	£100,000	£100,000	7/9	£38,750
Buffelsdoorn	550,000	500,000	119/-	2,975,000
Blinkpoort	50,000	38,000	16/3	30,875
Cornucopia	100,000	100,000	2/1	10,417
Eastleigh Mines	250,000	240,000	32/6	390,000
Elandslaagte	120,000	107,500	9/6	51,062
East Orion	275,000	275,000	14/-	192,500
Gordon Estate	175,000	160,000	14/4	114,667
Great Britain	72,100	60,100	*30/-	90,150
Leeuwpoort	100,000	95,000	3/6	16,625
Marais Reef	275,000	275,000	4/-	55,000
Meyer and Leeb	15,000	15,000	*7/6	5,625
Minerva	200,000	150,000	39/6	296,250
Molyneux Mines	50,000	37,500	42/6	79,687
New Florida	125,000	105,000	17/9	93,187
New Rietfontein	220,000	220,000	74/-	814,000
Nigel	160,000	160,000	145/-	1,160,000
New Midas Estate	100,000	100,000	40/6	202,500
Orion	160,000	160,000	*65/-	520,000
Roodepoort Heidelberg	160,000	105,522	*35/-	184,663
Rietkuil	120,000	90,000	*20/-	90,000
Steyn Estate	140,000	125,000	32/6	203,125
Vulcan	95,000	90,000	5/6	24,750
	£3,612,100	£3,308,622	...	£7,638,833

* Approximate Price.

SUMMARY.

	Nominal Capital.	Issued Capital.	Market Valuation. 19th June.
(*a*)	£15,001,000	£14,332,755	£70,298,399
(*b*)	1,547,000	1,343,000	4,032,875
(*c*)	3,612,100	3,308,622	7,638,833
Totals	£20,160,100	£18,984,377	£81,970,107

WITWATERSRAND GOLD MILLS.

Showing Stamps now running June 1895, and the Probable Number Five Years hence.*

Name of Company.	Stamps now running.	Probable Number 5 yrs. hence.	Name of Company.	Stamps now running.	Probable Number 5 yrs. hence.
Langlaagte Estate	160	160	Stanhope	20	...
New Primrose	160	200	Worcester	20	40
City and Suburban	130	240	Meyer and Leeb	15	15
Crown Reef	120	120	Ginsberg	10	30
Geldenhuis Estate	120	140	Lancaster	10	10
Jumpers	100	100	New Midas	10	10
Simmer and Jack	100	280	South Simmer	...	280
Langlaagte Royal	80	120	Simmer West	...	200
May Consolidated	80	100	Simmer East	...	200
Langlaagte Block B.	75	75	Central Nigel Deep	...	200
Glencairn	70	100	Crown Deep	...	200
Robinson	70	150	Durban Roodepoort Deep	...	200
New Kleinfontein	65	100	Geldenhuis Deep	...	200
New Crœsus	60	100	Glen Deep	...	200
Durban Roodepoort	60	70	Jumpers Deep	...	200
Langlaagte United	60	60	Knight's Deep	...	200
New Heriot	60	100	Langlaagte Deep	...	200
Randfontein	60	120	Nigel Deep	...	200
Jubilee	55	50	Nourse Deep	...	200
George Goch	60	80	Robinson Deep	...	200
Metropolitan	52	60	Rose Deep	...	200
Meyer and Charlton	50	50	Roodepoort Deep	...	150
New Rietfontein	50	100	Witwatersrand	...	120
Paarl Central	50	60	Champ d'Or Deep	...	100
United Main	50	90	Consolidated Angle-Tharsis	...	100
Van Ryn Estate	50	60	Ferreira Deep	...	100
Wemmer	50	70	Aurora West United	...	100
Champ d'Or	50	50	Village Main Reef	...	100
New Chimes	40	40	Horsham Monitor	...	80
Comet	40 }		Vogelstruis	...	80
Agnes Munro	... }		Balmoral	...	60
Cinderella	... }	240	Bonanza	...	60
St. Angelo	... }		North Randfontein	...	60
Blue Sky	... }		Langlaagte Star	...	60
Ferreira	40	80	Modderfontein	...	60
Henry Nourse	40	40	Unified	...	60
Orion	40	80	Driefontein	...	50
Wolhuter	40	140	Spes Bona	...	50
Salisbury	40	50	Minerva	...	40
Geldenhuis Main Reef	30	30	Treasury	...	40
Nigel	30	40	Van Ryn Western	...	40
Princess	30	60	Heidelberg Roodepoort	...	30
Banket	20	20	Main Reef	...	30
Johannesburg Pioneer	20	...			

Summary.

Stamps now running	2642
Probable number 5 years hence	8580

* In regard to the probable number of stamps running 5 years hence, our estimate will no doubt be considered rather conservative, but we prefer to err on this side rather than incur the risk of over-estimating.

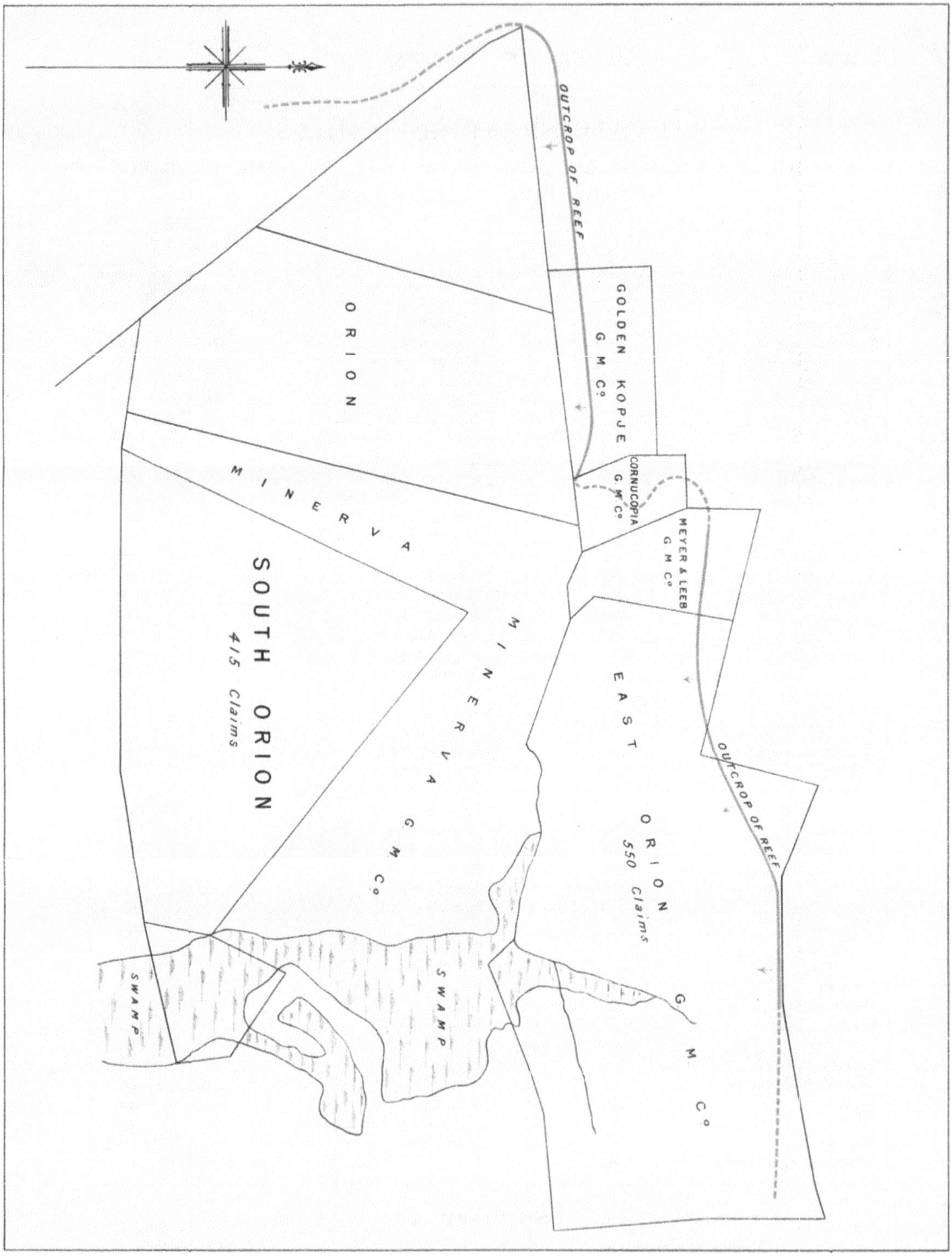

PLAN SHOWING SOME OF THE MORE IMPORTANT BLACK REEF PROPERTIES.

The material originally positioned here is too large for reproduction in this reissue. A PDF can be downloaded from the web address given on page iv of this book, by clicking on 'Resources Available'.

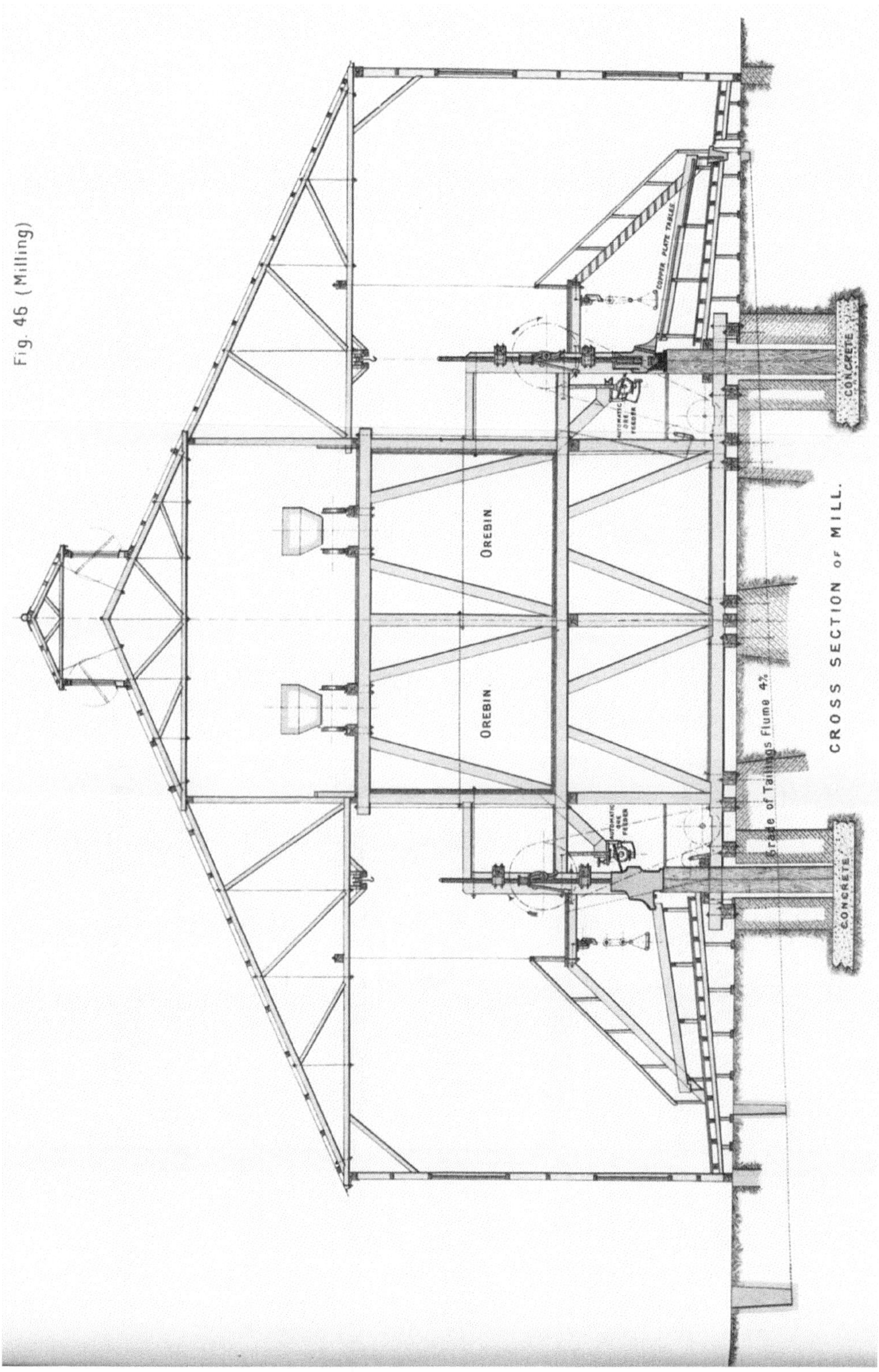

Fig. 46 (Milling)

The material originally positioned here is too large for reproduction in this reissue. A PDF can be downloaded from the web address given on page iv of this book, by clicking on 'Resources Available'.

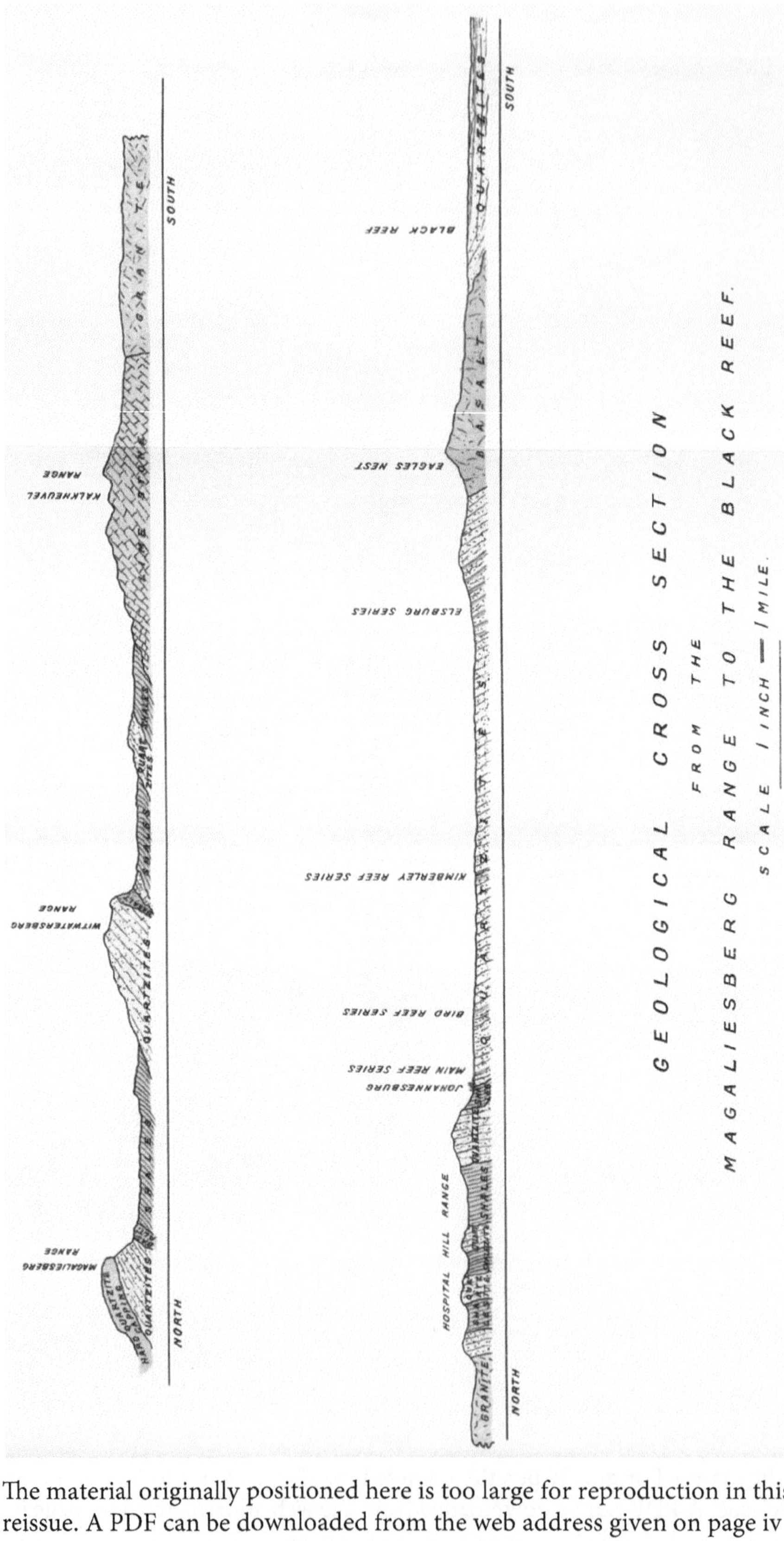

The material originally positioned here is too large for reproduction in this reissue. A PDF can be downloaded from the web address given on page iv of this book, by clicking on 'Resources Available'.

PLAN OF THE NIGEL PROPERTIES. *To accompany F.H.HATCH and J.A CHALMERS' "The Gold Mines of the Rand."*

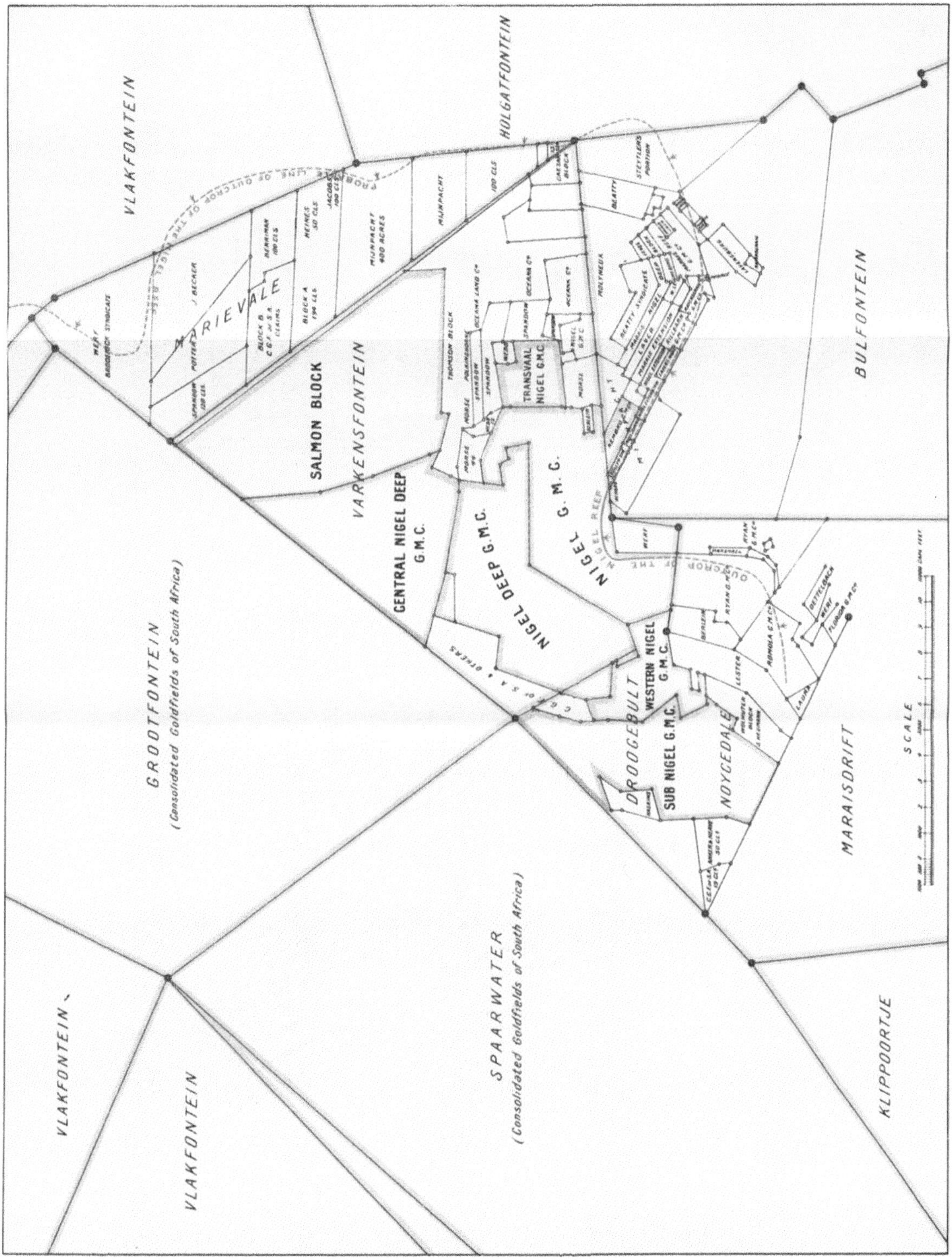

The material originally positioned here is too large for reproduction in this reissue. A PDF can be downloaded from the web address given on page iv of this book, by clicking on 'Resources Available'.

EXAMPLE OF A MINE PLAN AND SECTIONS.

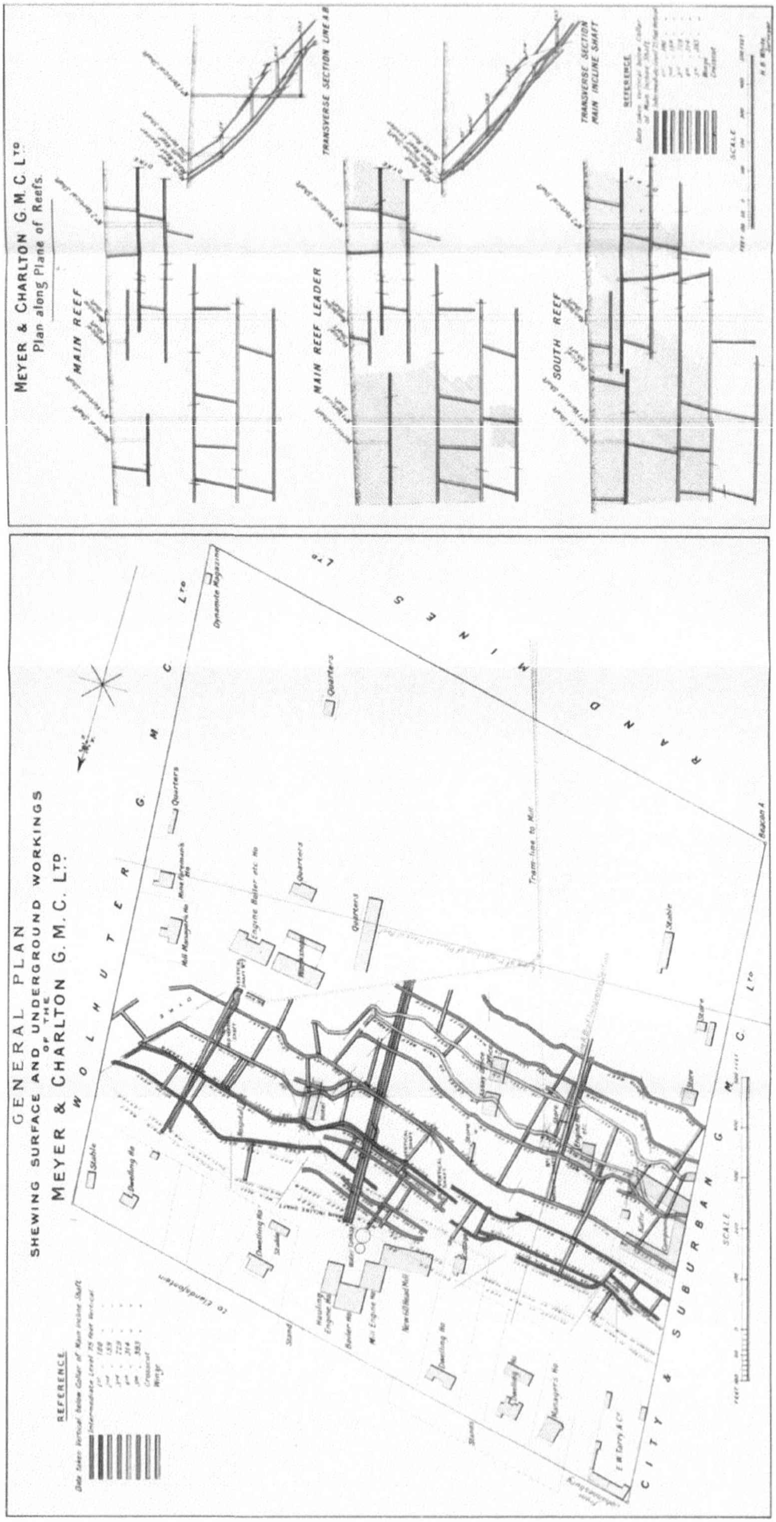

The material originally positioned here is too large for reproduction in this reissue. A PDF can be downloaded from the web address given on page iv of this book, by clicking on 'Resources Available'.

INDEX

THE END

Printed by R. & R. CLARK, LIMITED, *Edinburgh.*

GOLD MINES OF THE RAND

Correction for page 293

N.B.—Robinson shares are issued at £5. The market valuation of the shares of this company therefore should read £4,950,000